'This is a must read for anyone concerned with the workings of 21st-century fear culture and the media. Hammond provides an excellent account of how climate change has been integrated into a therapeutic outlook on public life.'

— *Frank Furedi, Emeritus Professor of Sociology, University of Kent, UK*

'Is climate change the ultimate global problem which must be solved through coordinated, collective action; or is this drive for consensus on climate change denying political subjectivity and obstructing the possibility of solving more pressing social injustices? Philip Hammond's new book, *Climate Change and Post-Political Communication*, offers an original and persuasive answer to this question by following the roles of media, news journalism and celebrity in the public framing of climate change.'

— *Mike Hulme, Professor of Human Geography, University of Cambridge, UK*

CLIMATE CHANGE AND POST-POLITICAL COMMUNICATION

For many years, the objective of environmental campaigners was to push climate change on to the agenda of political leaders and to encourage media attention to the issue. By the first decade of the twenty-first century, it appeared that their efforts had been spectacularly successful. Yet just at the moment when the campaigners' goals were being achieved, it seemed that the idea of getting the issue into mainstream discussion had been mistaken all along; that the consensus-building approach produced little or no meaningful action. That is the problem of climate change as a 'post-political' issue, which is the subject of this book.

Examining how climate change is communicated in politics, news media and celebrity culture, *Climate Change and Post-Political Communication* explores how the issue has been taken up by elites as potentially offering a sense of purpose or mission in the absence of political visions of the future, and considers the ways in which it provides a focus for much broader anxieties about a loss of modernist political agency and meaning. Drawing on a wide range of literature and case studies, and taking a critical and contextual approach to the analysis of climate change communication, this book will be a valuable resource for students and scholars of environmental studies, communication studies, and media and film studies.

Philip Hammond is Professor of Media and Communications at London South Bank University, UK.

ROUTLEDGE STUDIES IN ENVIRONMENTAL COMMUNICATION AND MEDIA

CLIMATE CHANGE AND POST-POLITICAL COMMUNICATION

Media, Emotion and Environmental Advocacy

Philip Hammond

Routledge
Taylor & Francis Group

LONDON AND NEW YORK

earthscan
from Routledge

First published 2018
by Routledge
2 Park Square, Milton Park, Abingdon, Oxon OX14 4RN

and by Routledge
711 Third Avenue, New York, NY 10017

Routledge is an imprint of the Taylor & Francis Group, an informa business

British Library Cataloguing-in-Publication Data
A catalogue record for this book is available from the British Library

Library of Congress Cataloging-in-Publication Data
A catalog record for this book has been requested

ISBN: 978-1-138-77749-1 (hbk)
ISBN: 978-1-138-77750-7 (pbk)
ISBN: 978-1-315-77259-2 (ebk)

Typeset in Bembo
by Deanta Global Publishing Services, Chennai, India

Printed in the United Kingdom
by Henry Ling Limited

CONTENTS

ACKNOWLEDGEMENTS

I am grateful to Dr Hugh Ortega Breton, with whom I co-authored a number of essays and conference papers on environmentalism and film, for many interesting conversations about politics and emotion in relation to climate change and other issues.

My thanks also to another long-term collaborator, Dr Andrew Calcutt, and to my colleague Prof. James Woudhuysen, for their insightful comments on the draft manuscript. The book's remaining faults result from me sometimes failing to follow their consistently sound advice.

Sincere thanks to my editor at Routledge, Annabelle Harris, and all her team, for their encouragement and forbearance.

As always, my biggest debt of gratitude is to my family for their tolerance of my often disruptive work schedule.

INTRODUCTION

'Post-political' climate change

> We have only just begun to understand what the idea of climate change is doing to us. Not just what it is doing to the ecological and physical conditions of our existence but, more importantly, to our political discourses, social relationships and imaginative worlds.
>
> (Hulme 2010: 273–4)

A strange thing has happened to the issue of climate change. For many years, it seemed that the objective of environmental campaigners was to push climate change on to the agenda of political leaders and to encourage media attention to the problem. The aim seemed to be to get everyone to agree that climate change was an important matter for public concern, and that action should be taken. It seemed to work. The politicians and journalists did pay attention, and (almost) everyone did agree that it was indeed very important. I watched this change happen at the university where I work. When I started here in the late 1990s, environmental concerns were not part of official university policy and nobody thought that was odd or unusual. Over the years, things gradually changed. Every office got a recycling bin. Low-energy lighting was installed, and turned off automatically if sensors detected no movement. When the university commissioned new buildings, they were designed on eco-friendly principles. Posters about sustainability appeared around the corridors and we held an exhibition about it in the courtyard. The environment became an important focus, not just of academic groups and research projects, but of corporate identity and institutional governance. Now, environmental concerns are very prominently part of official university policy and nobody thinks that is odd or unusual. A new consensus has been established; common-sense assumptions have shifted, not only in one university, but across society. Yet the strange thing is that just when everyone started to agree that climate change was a really important issue, it began to look as if that very agreement itself was a problem.

By the first decade of the twenty-first century, it appeared that the efforts of climate campaigners had been spectacularly successful. New energy policies were being put in place, new taxes and economic incentives were being introduced, new low-carbon lifestyles were being marketed and adopted, all in the name of reducing human impacts on the environment and combatting climate change. From presidents and prime ministers to rock stars and Hollywood actors, everyone was paying attention to the issue. As Erik Swyngedouw (2010: 215) describes, a consensus emerged that was 'now largely shared by most political elites from a variety of positions, business leaders, activists and the scientific community', with the 'few remaining sceptics...increasingly marginalized as either maverick hardliners or conservative bullies'. And yet, just at the moment when the campaigners' goals were being achieved, it started to seem that in fact this would not produce any meaningful change. Instead, we found ourselves in a 'paradoxical situation', in which a 'techno-managerial eco-consensus' insisted that 'we have to change radically, but within the contours of the existing state of the situation...so that nothing really has to change!' (Swyngedouw 2013: 2, 4). Similarly, Ingolfur Blühdorn and Ian Welsh (2007: 194) describe an early twenty-first century '*Zeitgeist* which insists on the official acknowledgement and incorporation of environmental concerns', but which at the same time continues the 'established practices and principles' of consumer capitalism.

Critics started to argue that the idea of getting the issue of climate change into mainstream discussion had been mistaken all along; that the consensus-building approach produced little or no meaningful action. As Amanda Machin (2013: 5) points out, 'Decisive action is underpinned not by consensus but by disagreement, for without a choice between real alternatives there can be no decision'. Yet the issue of climate change has come to be characterised by a 'strange non-political politics' that works through 'compromise, managerial and technical arrangement, and the production of consensus' (Swyngedouw 2013: 1, 5). That is the problem of climate change as a 'post-political' issue, which is the subject of this book.

The 'strange non-political politics' of climate change have been understood as symptomatic of a broader shift towards post-politics or post-democracy, and also as contributing to or helping to cause that shift. In both respects – as symptom and cause – this raises questions about whether part of the problem might be how climate change is represented, discussed and framed as an issue, and whether a different discursive framing might be a way to politicise it rather than reinforcing depoliticisation. With regard to media and popular culture, the role of emotions – particularly fear – is often understood to be key: apocalyptic depictions of climate change are supposed to prompt urgent action, but may instead result in paralysing dread. By way of introduction, let us consider each of these in turn: the concept of 'post-politics' that provides the wider context for this discussion; the apparently paradoxical character of contemporary climate change politics; and the role of media representation and framing in (de-)politicisation.

Post-politics?

Ideas about 'post-politics' or 'post-democracy' are attempts to understand what has happened to political life in Western societies since the end of the Cold War. As Slavoj Žižek describes it, there has been a 'degeneration of the political', in which former conflicts have been replaced by 'postmodern *post-politics*':

> In post-politics, the conflict of global ideological visions embodied in different parties which compete for power is replaced by the collaboration of enlightened technocrats (economists, public opinion specialists…) and liberal multiculturalists; via the process of negotiation of interests, a compromise is reached in the guise of a more or less universal consensus. Post-politics thus emphasises the need to leave old ideological divisions behind and confront new issues, armed with the necessary expert knowledge and free deliberation that takes people's concrete needs and demands into account.
>
> (Žižek 2000: 198)

With the end of the old ideological dividing lines of the Cold War era, politics has been hollowed out: we still use the terms 'Left' and 'Right' but these no longer relate to distinct worldviews or grand narratives of history. Capitalism has become the limit point and horizon of the possible, and the scope of political debate has shrunk to minor disagreements over technical matters of administration and management. Traditional differences between political parties seem less visible and less important, and established parties are sometimes rejected entirely by voters, in favour of 'outsider' figures or populist causes. Explosions of protest and anger rarely align with traditional patterns of political representation, but nor do they readily translate into stable new political formations, and the overall trend has been for people to disengage from formal politics and the public sphere.

As Žižek acknowledges, in a sense Francis Fukuyama's announcement of the 'End of History' at the end of the Cold War was accurate. While rejecting Fukuyama's triumphalist 'renaturalization of capitalism' (Žižek 2008: 405), radical critics of the post-political confirm his more downbeat assessment that we can now anticipate, not a bright and optimistic future, but 'centuries of boredom':

> The end of history will be a very sad time…the worldwide ideological struggle that called forth daring, courage, imagination, and idealism, will be replaced by economic calculation, the endless solving of technical problems, environmental concerns, and the satisfaction of sophisticated consumer demands.
>
> (Fukuyama 1989)

This has now become a common description of today's diminished political landscape. What Žižek and Fukuyama render in the framework of high-octane Hegelianism, Colin Crouch describes in empirical sociological terms with his idea of 'post-democracy'. This is not the same as the concept of the post-political, in that

Crouch is concerned with structural changes in power and institutions, but it does result in a similar view of contemporary public life. As he explains, the concept of 'post-democracy' describes:

> situations when boredom, frustration and disillusion have settled in…when powerful minority interests have become far more active than the mass of ordinary people in making the political system work for them; where political elites have learned to manage and manipulate popular demands; where people have to be persuaded to vote by top-down publicity campaigns.
>
> (Crouch 2004: 19–20)

Crouch's account, and indeed Fukuyama's, usefully draw attention to the fact that the 'triumph' of capitalism and liberal democracy was not an unalloyed victory for Western political and economic elites. The lack of a meaningful framework for political engagement and contestation meant that they were cut off from their own societies and deprived of mission and purpose. Concerns about a 'democratic deficit' express this problem: a society governed via market mechanisms and expert management and administration is in many ways of course a godsend for elites who do not have to make concessions to workers or engage the demos; but at the same time, if the public does not participate, or if it rejects detached structures of technocratic governance, then those elites lack legitimacy. The repeated popular rejection of the European Union in referenda across several countries over a number of years is a prime example.

As far as Crouch (2004: 104) is concerned, the changes described as post-politics or post-democracy are 'inexorable' and it is 'impossible to see any major reversal of them'. Instead, the task is simply to 'learn to cope' with post-democracy, 'softening, amending, sometimes challenging it' (2004: 12). Others contend that nothing fundamental has really changed. John Urry (2011: 91), for example, maintains that politics is alive and well in the grassroots and on the internet, and dismisses the post-political critique as nostalgia for a 'golden age' that never existed. To Jodi Dean (2011), claims about post-politics are 'childishly petulant': effectively, the Left is saying 'If the game isn't played on our terms, we aren't going to play at all. We aren't even going to recognize that a game is being played'. While conceding that there is some descriptive accuracy in accounts of the shift towards 'consensus, administration, and technocracy', she argues that business and financial interests are still waging political battles and that Žižek's arguments are 'better read as a critique of the left', which 'accepts capitalism' and has 'conceded to the right on the terrain of the economy'. If the Left would just have enough self-belief to reclaim this terrain, she implies, then politics could return to normal.

However, these critiques arguably miss the point. There is a distinction between everyday politics and 'the political' – understood as an underlying 'non-foundational foundation' of human social life. This rather awkward formulation is necessary because the theoreticians of the post-political adopt the 'post-foundational' view that there is 'no essential ground to any social order' (Wilson and Swyngedouw 2014: 10). This takes on the Heideggerian distinction between the ontic (everyday being) and the ontological (the possibility of being itself), and uses it to distinguish between everyday politics, as a 'contingent and incomplete attempt to ground a particular set of power relations on an ultimately absent foundation'; and 'the political', understood as that very absence, which 'undermines the social order constructed upon it, and which holds open the possibility of radical change' (Wilson and Swyngedouw 2014: 10). Post-politics means the closing off of the possibility of change, although the political is understood differently by different theorists. For Chantal Mouffe (2005), it is a trans-historical notion of antagonism between enemies; for Jacques Rancière (1999) it is the assertion of equality; and for Žižek (2002) it is class struggle. The political however, is only accessible via actual political phenomena – so the emergence of depoliticised consensus (as around climate change) is understood as problematic in that it conceals, disavows or forecloses the fundamental antagonism of the political, which is denied expression in the world of everyday politics.

Hence, Žižek says that post-politics should be understood as a 'wrong ideological impression':

> We don't really live in such a world, but the existing universe presents itself as post-political in the sense that there is some kind of a basic social pact that elementary social decisions are no longer discussed as political decisions. They are turned into simple decisions of gesture and of administration.
>
> (in Deichmann *et al.* 2002: 3)

Or as Anneleen Kenis and Matthias Lievens (2015: 22) put it in relation to climate change, the 'political foundation of society…can be rendered invisible by depoliticised discourses', thereby closing off 'the possibility to contest and change its order'.

At the same time, though, Dean's accusation of petulance is surely right in the sense that discussions of the post-political do often come across as an excuse. The possibility of politics is said to be 'closed off', 'ruled out', 'shut down' and so on, as if this was something *done to* or imposed upon the Left and for which it bore no responsibility.[1] It is as if the losing side in an argument complained that the winners were foreclosing the possibility of further disagreement by pretending that we now all agreed. If one continues to disagree, then why not say so? Unless perhaps one has run out of convincing arguments. If, as Žižek (in Derbyshire 2009) says, the 'only real question' is 'was Fukuyama right or not?', then the key post-Cold War political question remains whether a genuine progressive politics can be revived, reclaimed or reinvented in order to question this 'post-political' universe.

The premise of this book is that politics most certainly does not carry on as before, and that it will be a challenge to reinvent it. Such a reinvention will obviously not happen while we aim only at 'softening' a configuration that we accept as inevitable, but nor will it happen by simply pretending that left-wing politics is alive and well but just waiting in the wings somewhere off-stage. The specific aspect of the question addressed in the pages that follow is thus not only how and why the issue of climate change works in depoliticising ways today, but also whether it could instead be made to work differently, as part of efforts at re-politicisation.

Climate change as a 'post-political' issue

Climate change has become part of this post-political consensus – indeed, for Swyngedouw (2013: 3) it is the 'emblematic case' and '*cause célèbre* of de-politicization'. Critics of the post-political do not all agree with one another, either in terms of their general theoretical orientation, or about the specifics of how climate change should be understood as a post-political issue (see Wilson and Swyngedouw 2014); but a concern with the mainstreaming and de-politicisation of climate change is the common starting point. This is not seen as a problem inherent to green politics as such, but as something that has happened to it, or around it, and which has compromised its former radicalism. Swyngedouw, for example, who has done more than anyone to develop an analysis of climate change as part of the post-political consensus, suggests that the mainstreaming of contemporary environmentalism has 'unfolded in parallel' with broader contextual changes which have 'evacuated dispute and disagreement from the spaces of public encounter', implying that there is no necessary connection between environmentalism and the emergence of post-politics, only that the latter imposes a 'particular framing of climate change' which 'forecloses (or at least attempts to do so) politicization' (2010: 215, 227). He also argues that green politics itself has been ready to compromise, noting the 'rapid transformation' of organisations such as Greenpeace and the German Green Party. Where once these organisations offered 'a politics of contestation, organized action, radical disagreement and developing visionary alternatives', he argues, they have now been integrated into 'stake-holder-based negotiation arrangements aimed at delivering a negotiated policy' (2010: 227–8).

The assumption is that ecology used to embody a radical anti-capitalist outlook, but that this has now been incorporated into the mainstream as part of the development of consensual post-politics, and that this process of depoliticisation has then been reinforced by the institutionalisation of climate governance arrangements. Much as Crouch (2004: 98) describes post-democratic transformations in terms of new techniques and institutions of neoliberal governance, so Swyngedouw (2010: 215) argues that the depoliticisation of environmentalism is 'institutionally choreographed' in new 'postdemocratic institutional configurations'. It is not just that 'radical dissent, critique and fundamental conflict' are marginalised, but this is then further reinforced via 'democratically disembedded' institutional arrangements 'like "the Kyoto Protocol"; "the Dublin Statement", the "Rio Summit", etc.' (2010: 227).

This all seems descriptively accurate. We might think, for example, of the way that international climate summits purport to decide 'the future we want' (Ki-Moon 2012), outside of any democratic debate about, or mechanism of accountability for, that future. Yet if, as Chris Methmann (2010: 345) argues, 'climate protection' has become an 'empty signifier', adopted as an 'important policy goal' not only by national governments and the United Nations but also by global economic institutions such as the World Trade Organisation, the International Monetary Fund, the World Bank and the Organisation for Economic Cooperation and Development, it is not immediately obvious why this would be the case. Why would it appear necessary to 'integrate climate protection into the global hegemonic order without changing the basic social structures of the world economy' (Methmann 2010: 345)? Why would it seem important to incorporate environmental concerns into 'hegemonic' frameworks of post-political consensus? According to Swyngedouw (2010: 219, 228), environmentalism now works as a 'new opium for the masses', playing the 'radically reactionary' role of forestalling 'the articulation of divergent, conflicting and alternative trajectories of future socio-environmental possibilities', and 'reproducing, if not solidifying, a liberal-capitalist order for which there seems to be no alternative'.

In making this argument, both Swyngedouw (2010: 219) and Žižek (2008: 439) follow the lead of Alain Badiou, who said in a 2007 interview that:

> the rise of the 'rights of Nature' is a contemporary form of the opium of the people. It is an only slightly camouflaged religion: the millenarian terror, concern for everything save the properly political destiny of people, new instruments for the control of everyday life, the obsession with hygiene, the fear of death and of catastrophes…it is a gigantic operation in the depoliticisation of subjects. Behind it there is the idea that with strict ecological obligations one can prevent the emerging countries from competing too rapidly with the established imperial powers. The pressure exercised on China, India and Brazil has only just begun.
>
> (Badiou in Feltham 2008: 139)

Only the last of these points, about economic competition, indicates why Western elites might have a material interest in adopting an environmentalist agenda. But this does not explain the larger argument about the ideological character of environmentalism as a 'camouflaged religion' and the 'control of everyday life'.

To see the problem, we can compare Badiou's point with the very similar argument made more than three decades earlier by Jean Baudrillard, who said in 1970 that environmentalism was 'a new "opium of the people"' (Baudrillard 1974). Baudrillard put forward a traditionally Marxist ideology-critique, whereby he compared environmentalism to a 'witch-hunt', in that it attempted to unite antagonistic social classes in a 'new crusade' against a mystified threat by 'shouting apocalypse'. 'Nothing better than a touch of ecology and catastrophe to unite the social classes', he remarked caustically. It is a powerful critique, yet it no longer seems plausible to explain the official adoption of environmentalism as a way to paper over political divisions, an elite strategy to silence dissent and suppress demands for social change that might otherwise erupt at any moment. In 1970, Baudrillard pointed to things like the May 1968 student revolt in France and the opposition to the Vietnam War in the US as signs of a 'potential crisis situation'. In that context, it made logical sense to critique ecology as an 'ideology that could remake the holy union of mankind, beyond class discrimination, beyond wars, beyond neo imperialistic conflicts' (Baudrillard 1974). Today, however, those sorts of struggles and conflicts are conspicuous by their absence – as, indeed, the notion of 'post-politics' points out.

If they face no significant political challenge, why would Western elites need this new pacifying ideology? What now calls forth this 'ideological support structure for securing the socio-political status quo' (Swyngedouw 2010: 223)?

According to some, it is climate change itself that presents an existential challenge to the capitalist order. Methmann (2010: 369), for example, argues that 'climate mainstreaming' serves the goal of 'sustaining capitalism': environmentalism has adopted a way of 'remedying the dislocatory effects of climate change for hegemonic structures without changing them'. A similar argument is developed by Blühdorn (2007: 269), who argues that the elite's 'ostentatious declaratory commitment to effective action' should not be interpreted as 'evidence of the political will and ability to address and resolve the problems of unsustainability', but should rather be seen as 'a societal strategy for sustaining the unsustainable'.

The implication that, in the absence of a political challenge to the established order, the challenge instead comes from the threat of climate change is reminiscent of the ideas of 'risk society' theorists Ulrich Beck and Anthony Giddens, for whom it is the 'manufactured uncertainty' of industrial modernity that now drives social and political transformation, with risks and hazards themselves acting as 'quasi-subjects' (Beck 1998: 19). Beck and Giddens take the risk society thesis in different directions as regards climate change. Beck's work acknowledges that 'In the name of indisputable facts portraying a bleak future for humanity, green politics has succeeded in de-politicizing political passions', but retains a hope that the global threat may yet drive us towards a positive outcome by issuing a 'cosmopolitan imperative' for greater cooperation (Beck 2010: 263, 258). Giddens (2009: 114), in contrast, finds that tackling climate change is best served by his own brand of Third Way 'radical centrism' – the epitome of the post-political (Žižek 2000: 198). Giddens (2009: 114) argues that climate change should be 'lifted out of a right–left context' and the 'usual party conflicts should be suspended or muted'. Blühdorn (2007: 261) takes another view, arguing that the threats of risk society have been successfully neutralised by 'technological optimism' and a 'positive identification with the established system of democratic consumer capitalism'. If that were true, though, and a techno-optimist consumerism was the dominant outlook, then it would again return us to the puzzle of why elites feel it necessary to embrace discourses of environmentalism and sustainability.

Blühdorn's most interesting suggestion in this respect is that we have entered an era of 'simulative politics'. This is a 'seemingly schizophrenic condition where citizens want politics to be no more than symbolic, but still complain about democratic deficits and "merely symbolic" politics' (Blühdorn 2007: 265). This 'seemingly nonsensical form of political communication' arises from a condition where:

> citizens articulate demands which they do not want to see seriously implemented…a condition where citizens expect – in the sense of both *want* and *anticipate* – that the government *does not* seriously implement the demands which they, nevertheless, continue to articulate.
>
> (Blühdorn 2007: 264–5, original emphasis)

Drawing on Baudrillard's concept of simulation, Blühdorn characterises contemporary political communication as 'show politics' or the 'performance' of politics, whereby the goal is to maintain the appearance of 'the vitality and viability of politics itself' (2007: 267). Concerns about 'sustainability', he notes, are expressed not only in relation to the environment, but in all sorts of areas – 'pensions systems, health care systems, transport systems, the system of representative democracy and so forth': all are seen as 'unsustainable'.

> In all of these policy areas there is a sense of acute crisis and much talk about 'radical shake-ups', 'tough decisions' and 'hard policy roads'. There is a striking consensus between political elites and general electorates that it is time to stop *talking* about things and take *decisive action*: Cut through the rhetoric! Get down to the issues!
>
> (Blühdorn 2007: 252, original emphasis)

Yet such urgent demands for action simply add another layer of performance, he argues: the 'performance of seriousness'. What motivates the performance is the post-political situation itself: the 'exhaustion of authentic politics' and the fact that 'there is no vision of any viable alternatives' (2007: 265). In the 1970s and 1980s, Blühdorn suggests, the 'anti-politics' of radical ecologists meant 'rebellion against established mainstream politics' and its replacement by 'the authentic politics that was supposedly being rehearsed in the societal margins'. Today, however, 'anti-politics' just means 'frustration with politics', 'disengagement' and 'depoliticisation' (2007: 261–2). In this situation, the performance of '*authentic* eco-politics' has itself 'become ideological', since it sustains the simulation of meaningful political engagement (2007: 269, original emphasis).

Most critics of the post-political do not share Blühdorn's pessimism about democratic politics, and indeed the point of the critique is to think about how the post-political might be challenged and overcome. Rather than seeing contemporary political communication in terms of simulation, others have sought to analyse how particular discursive framings of climate change depoliticise the issue, and to figure out how it might be represented differently.

Media, culture and emotion

If, as Kenis and Lievens (2014: 5) argue, 'Fundamentally, depoliticisation is situated on the level of representation', then we need to understand exactly how this works not only in political discourse but also across the media and popular culture. Much previous work on the media and environmental issues has assumed that political and media elites are reluctant to pay attention and need to be pressured and encouraged to do so – an assumption that no longer seems to make sense in the context of the post-political mainstreaming of environmental concerns. Regarding climate change specifically, most critics have understood the problem in terms of a lack of consensus, with the media giving too much space to a diversity of conflicting views about the causes of and responses to global warming (for example, Boykoff and Boykoff 2007). But what if the problem was the opposite? What if the promotion of consensus and narrowing of debate was encouraging inaction and disengagement rather than promoting change?

According to Anabela Carvalho (2010: 172), the news media play an important role in 'processes of political (dis)engagement in relation to climate change', in terms of the way that media representations 'construct particular "subject positions" for individuals and cultivate dispositions to action or inaction'. Drawing on research which has shown how the public is often depicted as passive and 'childlike', with its opinions represented in terms of 'moods' and 'emotions' rather than rational political perspectives (Lewis *et al.* 2004), Carvalho argues that the media generally present climate change as the concern of elite decision-makers, and relegate ordinary citizens to the role of 'spectators or bystanders' (Carvalho 2010: 174–6).

Carvalho's comments here usefully highlight a further important aspect of debates about the post-political: the emotional dimension of representations of climate change. The most obvious aspect of this is, as Swyngedouw (2010: 217) notes, the 'continuous invocation of fear and danger'. Contemporary representations of climate change frequently involve 'millennial fears' and 'apocalyptic rhetoric', encouraging a paralysing uncertainty in the face of 'an overwhelming, mind-boggling danger…that threatens to undermine the very coordinates of our everyday lives and routines' (Swyngedouw 2010: 218). This is undoubtedly true, and indeed is what one would expect in the context of broader development of risk consciousness and a 'culture of fear' (Furedi 2005) – in both politics (Lipschutz 1999) and the media (Altheide 2002) – at the turn of the twenty-first century. Some have argued that fear is a useful mobiliser for environmental action (McNeill Douglas 2010), but it is also often viewed as problematic for various reasons. Sometimes it is rejected simply because it is inaccurate, as in James Lovelock's acknowledgement that he was too alarmist about 'Gaia's revenge', for example (Johnston 2012). More often, what the Institute for Public Policy Research (IPPR) think-tank describes as an 'alarmist repertoire', characterized by an 'inflated or extreme lexicon', an 'urgent tone and cinematic codes', and a 'quasi-religious register of death and doom' is seen as ineffective or even counter-productive since it offers only a 'counsel of despair' (Ereaut and Segnit 2006: 7). Such criticisms have not meant that apocalyptic, fearful constructions of climate change no longer feature prominently in public discourse: as the IPPR suggests, the 'sensationalism and connection with the unreality of Hollywood films' that such apocalypticism involves may be 'secretly thrilling'; a form of 'climate porn' (2006: 7).

While much of this discussion stays at the level of a pragmatic assessment of whether evoking fear is useful or not, there are other, potentially more interesting things to say about the emotional dimension of climate change communication. For one thing, it is striking that the importance of emotion and affect has repeatedly been rediscovered. In a 2016 survey of the field, Susanne Moser noted that greater attention was being given to 'the role of emotions in climate change communication', explaining that:

This greater focus on the affective and emotional (as opposed to just the cognitive) side of climate change is partly driven by the irrational-seeming lack of concern about the problem and persistent psychological distancing, partly by the often intense emotional reactance to climate change (and its messengers) by those who do not 'believe' in climate change, and partly by the increasingly observed sense of despair and hopelessness among those who understand the science and experience early impacts and/or the lack of commensurate action.

(Moser 2016: 6)

Moser was updating an earlier, 2010 survey, in which she had noted how researchers were beginning to realise that 'messages are more than the words or information conveyed' and that attention needed to be paid to 'the emotions that are being evoked' (Moser 2010: 40). Birgitta Höijer (2010: 718), for example, published an interesting study of how news reports 'address a number of different emotions to anchor climate change in everyday discourse'. In 2009 – the year, incidentally, that the academic journal *Ecopsychology* was established[2] – a report by the American Psychological Association's Task Force on the Interface Between Psychology and Global Climate Change noted that 'risk perceptions…are influenced by associative and affect-driven processes as much or more than by analytic processes' (Swim *et al.* 2009: 37). The implication that emotional appeals are as important, or more important, than information and rational argument had already been made explicitly in a number of studies. Anthony Leiserowitz (2006: 63) had argued that 'risk perception is greatly influenced by affective and emotional factors', for instance; while Elke Weber (2006: 116) had suggested that 'attention-catching and emotionally-engaging informational interventions may be required to engender the public concern necessary for individual or collective action in response to global warming'.

In fact even then, the idea that affect is more important than cognitive reasoning was not a new one. There was already a long research tradition in psychology making this point both in general terms and specifically in relation to environmentalism. Julie Ann Pooley and Moira O'Connor, for instance, argued at the turn of the century that 'the key entry point for environmental education is via the affective domain' (Pooley and O'Connor 2000: 712). It may be that this argument – for the importance, even the primacy, of emotional reactions over logical thought – continues to be reiterated (often as if it were a new insight) because it chimes with contemporary concerns and assumptions. In relation to environmental campaigning, it offers an explanation for why information campaigns do not appear to be entirely successful in persuading people to 'go green' (Ramesh 2011). More broadly in politics in recent years we have seen the rise of the idea of 'nudging' the public to adopt particular ideas or behaviours rather than making an argument that they should do so (Thaler and Sunstein 2008). The notion that political leaders should 'nudge' citizens, or that they should seek to address voters' emotional wellbeing (Richards 2007), perhaps reflects the difficulty that such leaders and other political actors now have in making directly *political* arguments, given the exhaustion of Left/Right politics since the late 1980s.

The loss of the modernist Left/Right framework has provided fertile ground for the development, not only of a 'culture of fear', but also of what has been referred to as 'post-traumatic culture' (Farrell 1998), 'traumaculture' (Luckhurst 2003), or 'therapy culture' (Furedi 2004). This emotional, therapeutic discourse provides a framework for managing individual anxieties about agency in a social situation characterised by greater uncertainty and a perception of powerlessness. This implies that, while Swyngedouw (2010: 218–9) and others are no doubt correct to argue that visions of eco-apocalypse are both symptomatic of, and also reinforce, a post-political outlook in which the urgency of impending climate catastrophe closes down democratic debate about possible futures, this does not work through fear alone. Apocalyptic framings of climate change give indirect expression to the loss of modernist political subjectivity, but not simply through negation: rather, agency is re-interpreted in moral and therapeutic terms relating to how people should conduct themselves in relation to the environment. Hence, as Amanda Rohloff (2011: 640) says of Al Gore's high-profile environmental campaigning, notwithstanding the presentation of climate change in apocalyptic terms, a de-politicised presentation of the issue as a 'moral' question actually provides a kind of assurance: 'global warming becomes not an uncertain risk, but a moral certainty'. This does not suggest any break from a post-political framework, but it does indicate that there is more going on than simply scaring people.

A frequent observation about environmental apocalypticism is that, unlike in religious conceptions, it involves no moment of transcendence or redemption (Swyngedouw 2010: 218–9, Levene 2010: 60). As Pascal Bruckner puts it:

> The Christian Apocalypse presented itself as a revelation, a passage in to another temporal order, whereas this apocalypse reveals nothing, it issues the final judgment: pure apocalypse. No promise of redemption, just an ideal for survivors, an 'epidemic of remorse'.
>
> (Bruckner 2013: 65)

Yet rather than a complete absence of redemption or salvation, it might be more accurate to say that environmental apocalyptic narratives offer a kind of pseudo-redemption: there is a sense of individuated, personalised, therapeutic redemption, to be achieved through inward-focussed projects of the self. This therapeutic appeal explains why emotional engagement continues to be offered as the extra magic ingredient; the thing that could potentially make both media and celebrity campaigning work effectively (Doyle *et al.*, forthcoming), and that could even address the wider problem of political disengagement and the democratic deficit (Richards 2004).

About this book

Many books on climate change communication, perhaps close to all of them, are basically concerned with making such communication more effective, whether that is understood in terms of the transmission of scientific knowledge or the dynamics of public engagement. This is not one of those books. Given the way that, as critics of the post-political point out, environmentalism has become mainstreamed in recent decades, it would be limiting to adopt this sort of pragmatic, policy-oriented approach. Moreover, I have consciously tried not to take any of the assumptions in the literature for granted, mainly because the situation seems to demand it: what many people used to take for granted regarding the oppositional, radical anti-capitalist character of green politics and the reluctance of elites to engage with the issue of climate change no longer seems to describe contemporary experience. If we keep expecting or wanting things to conform to these assumptions we are unlikely to see the present for what it is. That is the reason why the book works with the concept of the 'post-political'.

In the spirit of not taking established assumptions on trust, the first two chapters review the history of how environmental issues, particularly climate change, have been taken up in mainstream politics and news reporting. Chapter 1 tries to make sense of the way that climate change emerged as a political issue in the late 1980s by putting this in the context of both the prior history of elite engagement with environmental issues and the particular challenges that political leaders faced at the end of the Cold War. As this implies, the incorporation or mainstreaming of environmental concerns is not a recent phenomenon: it has happened before, at the very moment, in fact, that is now seen as the high point of ecological radicalism, the early 1970s. Back then, the issue worked for elites in much the way described by Baudrillard at the time, as a means to cohere a divided society. But environmental concerns were also taken up by establishment politicians as a potential source of meaning or spiritual renewal: a way to offer their societies some higher purpose than mere consumer satisfaction. The end of the Cold War encouraged a new generation of political leaders to seize on those same possibilities, finding meaning and purpose in the fight against climate change in order to help fill the void created by collapse of Left/Right politics. This is the Fukuyamian moment so central to Žižek's argument about the emergence of post-politics, of course, when the ending of modernist grand narratives of history made the challenge of discovering a new political vocabulary both more urgent and more difficult to accomplish. The result was the assimilation of more genuinely ecological themes into mainstream political discourse, and also the emergence of an emotional, therapeutic framing.

Chapter 2 offers a meta-analysis of research on the ups and downs of the environment as a news issue over time, interrogating the different ways that such fluctuations have been theorised and explained. These patterns have been widely misunderstood, it is argued, mainly because critics' assumptions about the media's attitude to reporting on climate change have led them to misinterpret the evidence. The result is that much work in this area not only adopts an incoherent model of the relationship between media and centres of political power, but also works to enforce conformity and consensus in an uncritical fashion.

While Chapter 2 focuses on the news media in their reporting role, Chapter 3 examines the role of the media as campaigners in their own right, trying to raise public awareness of environmental concerns and to advocate for change. Analysts of the post-political have criticised media research agendas which simply encourage greater consensus in news reporting, and have instead argued for more attention to be paid to the ways in which the media may either promote depoliticisation or work to politicise the issue of climate change. Here we follow this newer agenda, looking in detail at two contrasting case studies, both of which involve eco-friendly 'lifestyle journalism', but in different ways – one example, the BBC, is constrained by conventions of journalistic impartiality and could be said to have an institutional obligation to depoliticise; whereas the other example, the *Guardian*, is regarded (and regards itself) as a leading oppositional voice on environmental issues. A dilemma of green lifestyle journalism is that by focussing on consumer behaviour it risks inadvertently promoting the very thing it seeks to critique. Yet here too, it is argued, the problem has not been well understood: what this genre of journalism promotes is not consumerism so much as an etiquette of 'correct' behaviour and self-monitoring.

Somewhat similar concerns are the starting point for Chapter 4, which looks at the phenomenon of celebrity environmental advocacy, including the efforts both of activist *celebrities* such as Leonardo DiCaprio, and of celebrity *activists* such as Al Gore. The domain of celebrity again involves a potentially jarring juxtaposition of anti-consumerist campaigning with luxury consumer lifestyles, of course. Celebrity campaigners have been roundly disparaged as offering 'heroic' role models and promoting their own 'brand' as much as the causes they espouse, but they have also been praised for their potential to act as 'emotional pedagogues', connecting with audiences in a more visceral way and teaching us how to feel about climate change. Assessing these debates, the chapter argues that rather than the 'politicisation of emotion' that some critics have seen in recent celebrity interventions, what is actually on offer is an emotionalisation of politics, translating issues of public concern into personal, therapeutic terms.

Chapter 5 extends this discussion, re-examining the relationship between celebrities and audiences, and the sorts of solutions to climate change that celebrity campaigners promote. Counter-intuitively, recent research has suggested that the key audience for celebrity campaigning is not the public but elites, for whom celebrity engagement with an issue stands in for public interest in it. In this respect, the involvement of celebrities in climate change campaigning is symptomatic of 'post-democratic' and 'simulative' forms of political communication. The policy proposals typically offered in celebrity interventions have often been criticised as not challenging, or even addressing, the systemic dimension of the problem of climate change, focusing instead on small-scale, individualistic, light-bulb-changing actions. The chapter concludes with an analysis of Naomi Klein's more radical approach, which takes on many of the conventions of celebrity emotional advocacy but seeks to avoid the pitfalls identified by critics of celebrity campaigning. While it certainly cannot be accused of adapting to a techno-managerialist outlook, however, it seems that the greater radicalism of Klein's perspective also means greater anti-modernism, cutting away the ground of the political agency it ostensibly seeks to promote.

As may be apparent already, there is a line of argument running through the book about where to draw the line between the mainstream and the marginal, the consensus and critique. This is what makes the phenomenon of 'post-political' climate change an interesting one: it demonstrates quite clearly that the ground has shifted and that we need to rethink critique. We appear to be offered a choice between a techno-managerial administrative consensus on one side, and a repudiation of modernist subjectivity on the other. Most fundamentally, we need to rethink whether climate change has any place in progressive politics, an issue to which we return in the Conclusion.

Notes

1   It is also worth noting that, as Swyngedouw and Wilson (2014: 302) point out, 'the original problem of depoliticisation addressed by post-foundational theory was not post-politics, but the economic determinism of orthodox Marxism that had stifled the intellectual freedom of the Left'. This is one way of reading it, perhaps. Another would be to say that radicals were eager to jettison the working class as the political subject of historical change because that was preferable to admitting their own political failure. The paradigmatic text in this tradition is Laclau and Mouffe's 1985 work *Hegemony and Socialist Strategy*.
2   See http://www.liebertpub.com/overview/ecopsychology/300/. This is a US publication. A *European Journal of Ecopsychology* (http://eje.naturalresourceswellbeing.com) was also founded the following year.

References

Altheide, David L. (2002) *Creating Fear: News and the Construction of Crisis*. New York, NY: Aldine de Gruyter.

Baudrillard, Jean (1974) The Environmental Witch-Hunt. Statement by the French Group 1970, in Reyner Banham (ed.) *The Aspen Papers: Twenty Years of Design Theory from the International Design Conference in Aspen*. New York, NY: Praeger [available at www.metamute.org/editorial/articles/environmental-witch-hunt].

Beck, Ulrich (1998) Politics of Risk Society, in Jane Franklin (ed.) *The Politics of Risk Society*. Cambridge: Polity.

Beck, Ulrich (2010) Climate for change, or how to create a green modernity? *Theory, Culture & Society*, 27 (2–3): 254–66.

Blühdorn, Ingolfur (2007) Sustaining the unsustainable: Symbolic politics and the politics of simulation, *Environmental Politics*, 16 (2): 251–75.

Blühdorn, Ingolfur and Ian Welsh (2007) Eco-politics beyond the paradigm of sustainability: A conceptual framework and research agenda, *Environmental Politics*, 16 (2): 185–205.

Boykoff, Maxwell T. and Jules M. Boykoff (2007) Climate change and journalistic norms: A case-study of US mass-media coverage, *Geoforum*, 38 (6): 1190–1204.

Bruckner, Pascal (2013) *The Fanaticism of the Apocalypse*. Cambridge: Polity.

Carvalho, Anabela (2010) Media(ted) discourses and climate change: A focus on political subjectivity and (dis)engagement, *WIREs Climate Change*, 1 (2): 172–9.

Crouch, Colin (2004) *Post-Democracy*. Cambridge: Polity.

Dean, Jodi (2011) Post-politics? No, thanks! *Future Non Stop*, http://future-nonstop.org/c/b122b85eff80835dfd654453d325ba0b.

Deichmann,Thomas, Sabine Reul and Slavoj Žižek (2002) About war and the missing centre in politics, *Eurozine*, 15 March, www.eurozine.com/about-war-and-the-missing-center-in-politics/?pdf.

Derbyshire,Jonathan (2009) Interview with Slavoj Žižek, *New Statesman*, 29 October, www.newstatesman.com/ideas/2009/10/today-interview-capitalism.

Doyle, Julie, Nathan Farrell and Michael K. Goodman (forthcoming) Celebrities and Climate Change: History, Politics and the Promise of Emotional Witness, in Matthew C. Nisbet (ed.) *The Oxford Encyclopedia of Climate Change Communication*. Oxford: Oxford University Press, http://climatescience.oxfordre.com/page/climate-change-communication/.

Ereaut, Gill and Nat Segnit (2006) *Warm Words: How Are We Telling the Climate Story and Can We Tell It Better?* London: Institute for Public Policy Research.

Farrell, Kirby (1998) *Post-Traumatic Culture: Injury and Interpretation in the Nineties*. Baltimore, MD: Johns Hopkins University Press.

Feltham, Oliver (2008) *Alain Badiou: Live Theory*. London: Continuum.

Fukuyama, Francis (1989) The end of history? *The National Interest*, Summer, www.wesjones.com/eoh.htm.

Furedi, Frank (2004) *Therapy Culture: Cultivating Vulnerability in an Uncertain Age*. London: Routledge.

Furedi, Frank (2005) *Culture of Fear: Risk Taking and the Morality of Low Expectations* (Revised Edition). London: Continuum.

Giddens, Anthony (2009) *The Politics of Climate Change*. Cambridge: Polity.

Höijer, Birgitta (2010) Emotional anchoring and objectification in the media reporting on climate change, *Public Understanding of Science*, 19 (6): 717–31.

Hulme, Mike (2010) Cosmopolitan climates, *Theory, Culture & Society*, 27 (2–3): 267–76.

Johnston, Ian (2012) 'Gaia' scientist James Lovelock: I was 'alarmist' about climate change, *msnbc.com*, 23 April, http://worldnews.msnbc.msn.com/_news/2012/04/23/11144098-gaia-scientist-james-lovelock-i-was-alarmist-about-climate-change.

Kenis, Anneleen and Matthias Lievens (2014) Searching for 'the political' in environmental politics, *Environmental Politics*, 23 (4): 531–48. DOI:10.1080/09644016.2013.870067 [accessed electronically].

Kenis, Anneleen and Matthias Lievens (2015) *The Limits of the Green Economy: From Reinventing Capitalism to Repoliticising the Present*. Abingdon: Routledge.

Ki-Moon, Ban (2012) The future we want, *New York Times*, 23 May, www.un.org/sg/articles/articleFull.asp?TID=128&Type=Op-Ed&h=1.

Laclau, Ernesto and Chantal Mouffe (1985) *Hegemony and Socialist Strategy*. London: Verso.

Leiserowitz, Anthony (2006) Climate change risk perception and policy preferences: The role of affect, imagery, and values, *Climatic Change*, 77 (1–2): 45–72.

Levene, Mark (2010) The Apocalyptic as Contemporary Dialectic: From Thanatos (Violence) to Eros (Transformation), in Stefan Skrimshire (ed.) *Future Ethics: Climate Change and Apocalyptic Imagination*. London: Continuum.

Lewis, Justin, Karin Wahl-Jorgensen and Sanna Inthorn (2004) Images of citizenship on television news: Constructing a passive public, *Journalism Studies*, 5 (2): 153–64.

Lipschutz, Ronnie D. (1999) Terror in the suites: Narratives of fear and the political economy of danger, *Global Society*, 13 (4): 411–39.

Luckhurst, Roger (2003) Traumaculture, *New Formations*, 50: 28–47.

Machin, Amanda (2013) *Negotiating Climate Change: Radical Democracy and the Illusion of Consensus*. London: Zed Books.

McNeill Douglas, Richard (2010) The Ultimate Paradigm Shift: Environmentalism as Antithesis to the Modern Paradigm of Progress, in Stefan Skrimshire (ed.) *Future Ethics: Climate Change and Apocalyptic Imagination*. London: Continuum.

Methmann, Chris (2010) Climate protection as empty signifier, *Millennium: Journal of International Studies*, 39 (2): 345–72.

Moser, Susanne C. (2010) Communicating climate change: History, challenges, process and future directions, *WIREs Climate Change*, 1 (1): 31–53.

Moser, Susanne C. (2016) Reflections on climate change communication research and practice in the second decade of the 21st century: What more is there to say? *WIREs Climate Change*, 7 (3) DOI:10.1002/wcc.403 [accessed electronically].

Mouffe, Chantal (2005) *On the Political*. Abingdon: Routledge.

Pooley, Julie Ann and Moira O'Connor (2000) Environmental education and attitudes: Emotions and beliefs are what is needed, *Environment and Behavior*, 32 (5): 711–23.

Ramesh, Randeep (2011) Public support for tackling climate change declines dramatically, *Guardian*, 7 December, www.guardian.co.uk/environment/2011/dec/07/public-support-climate-change-declines.

Rancière, Jacques (1999) *Disagreement: Politics and Philosophy*. Minneapolis, MN: University of Minnesota Press.

Richards, Barry (2004) The emotional deficit in political communication, *Political Communication*, 21 (3): 339–52.

Richards, Barry (2007) *Emotional Governance*. Basingstoke: Palgrave Macmillan.

Rohloff, Amanda (2011) Extending the concept of moral panic: Elias, climate change and civilization, *Sociology*, 45 (4) 634–49.

Swim, Janet, Susan Clayton, Thomas Doherty, Robert Gifford, George Howard, Joseph Reser, Paul Stern and Elke Weber (2009) *Psychology and Global Climate Change: Addressing a Multi-Faceted Phenomenon and Set of Challenges*. Report by the American Psychological Association's Task Force on the Interface Between Psychology and Global Climate Change. Washington, DC: American Psychological Association, www.apa.org/science/about/publications/climate-change.pdf.

Swyngedouw, Erik (2010) Apocalypse forever? Post-political populism and the spectre of climate change, *Theory, Culture & Society*, 27 (2–3): 213–32.

Swyngedouw, Erik (2013) The non-political politics of climate change, *ACME: An International E-Journal for Critical Geographies*, 12 (1): 1–8.

Swyngedouw, Erik and Japhy Wilson (2014) There Is No Alternative, in Japhy Wilson and Erik Swyngedouw (eds.) *The Post-Political and Its Discontents*. Edinburgh: Edinburgh University Press.

Thaler, Richard H. and Cass R. Sunstein (2008) *Nudge: Improving Decisions About Health, Wealth and Happiness*. New Haven, CT: Yale University Press.

Urry, John (2011) *Climate Change and Society*. Cambridge: Polity.

Weber, Elke U. (2006) Experience-based and description-based perceptions of long-term risk: Why global warming does not scare us (yet), *Climatic Change*, 77 (1–2): 103–20.

Wilson, Japhy and Erik Swyngedouw (2014) Seeds of Dystopia: Post-Politics and the Return of the Political, in Japhy Wilson and Erik Swyngedouw (eds.) *The Post-Political and Its Discontents*. Edinburgh: Edinburgh University Press.

Žižek, Slavoj (2000) *The Ticklish Subject: The Absent Centre of Political Ontology*. London: Verso.

Žižek, Slavoj (2002) *For They Know Not What They Do: Enjoyment as a Political Factor* (Second Edition). London: Verso.

Žižek, Slavoj (2008) *In Defense of Lost Causes*. London: Verso.

1

POLITICAL ELITES AND THE SEARCH FOR GREEN MEANING

This chapter focuses on the political mainstream, examining how Anglo-American politicians have engaged with environmentalism, particularly around the issue of climate change. National political leaders are of course understood to be central to attempts to address environmental problems. There is, however, an ambiguity about the relationship between political elites and the public on the issue of climate change. Environmental campaigners and activists often tend to assume that things work bottom-up: they build public support in order to get politicians to act on environmental concerns. This was the rationale for the September 2014 People's Climate March, for example, which was designed to put pressure on world leaders gathering at a UN climate summit in New York, on the grounds that governments would not act without popular demand from below.[1] Academic studies also often assume that 'governments are unlikely to [act] without public pressure' from a 'large, well-orchestrated and sustained climate movement' (Roser-Renouf et al. 2014: 163). At the same time, though, climate change is also often implicitly understood as an elite concern, about which the public needs to be educated and roused to action via top-down initiatives. In *The Politics of Climate Change*, for instance, Anthony Giddens notes that public support for climate change policies is 'likely to wax and wane', and advises that governments must try to 'foster a more widespread consciousness of the need for action' (Giddens 2009: 230). An example of the sort of effort he has in mind might be the television advertising campaign launched in 2009 by the British government, at a cost of £6m, to raise public awareness about climate change. The campaign (discussed further below) apparently failed to strike a chord with the public: it later featured on a list, compiled by the UK Advertising Standards Authority, of the top ten most complained about adverts of all time (Westcott 2012).

It is difficult to make sense of the dual role that the public are understood to play: as both the grassroots force that demands action from reluctant leaders, and the passive and unconcerned target of elite attempts to gain support for environmental policies. The take-up of climate change as an issue in mainstream politics is best understood, it is argued here, as an elite-driven project. Examining the context of its emergence at the end of the 1980s, the chapter highlights how the issue appealed to politicians because it was perceived as having the potential to provide a new source of political meaning for the post-ideological, post-Cold War world. The issue of climate change offered leaders a fresh way to conceive of political action as taking place on a large historical scale, and a new way to think about the future in a context in which the familiar framework of Left and Right was no longer viable. As we shall see, this necessarily entailed finding a new, post-political, emotional discourse with which to speak for an imagined constituency, including children and future generations. This discourse did not need to be invented from scratch, however: it was more a case of reworking themes which had already emerged in politicians' engagement with environmentalism in earlier contexts.

Although, as discussed in the Introduction, climate change can indeed be characterised as a post-political issue, it is nevertheless routinely discussed in ways that map it on to the old framework of Left/Right politics. Climate change is variously understood as a left-wing and progressive issue or sometimes as a conservative issue; as a divisive or a consensus issue; as party-political or bipartisan; as marginal to the mainstream or as part of it. The picture is complicated by the fact that a complex relationship between environmentalism and mainstream politics had developed over the two decades preceding the emergence of climate change as the central concern. We therefore begin by examining modern environmentalism's emergence within the framework of Left/Right politics.

Environmentalism in the age of Left and Right

Accounts of modern environmentalism's emergence usually date it to the decade between the publication of Rachel Carson's *Silent Spring* in 1962, and the US government's ban on DDT (the pesticide which was the target of her book) in 1972, but with a marked quickening toward the end of that period. Friends of the Earth was founded in 1969, the first Earth Day was held in 1970, Greenpeace was established in 1971, and 1972 saw the first United Nations environmental conference (held in Stockholm) as well as the publication of the Club of Rome's *The Limits to Growth* report and of *The Ecologist* magazine's *A Blueprint for Survival*. The rise of the modern environmental movement is often associated with other radical protest campaigns of the era, particularly the Vietnam anti-war movement (Haq and Paul 2012: 7), yet its concerns apparently became part of the mainstream very quickly – indeed, instantaneously. The Earth Day Network (n.d.) claims with some justification that the mass mobilisation for Earth Day 1970 marked 'what many consider the birth of the modern environmental movement', but according to Tarla Rai Peterson (2004: 19), 'By 1970…[the environmental movement] had become mainstream'. As evidence of this mainstreaming, Peterson points to the fact that President Richard Nixon addressed the issue of the environment in his 1970 State of the Union speech. It would appear, then, that environmentalism was born as a radical grassroots protest movement that was simultaneously part of mainstream politics.

One way to approach this puzzle would be through Andrew Dobson's (2007: 2–3) argument that there is a sharp division between 'environmentalism' and 'ecologism': the former is a 'managerial approach' which proposes no fundamental social or economic change, whereas the latter 'presupposes radical changes in our relationship with the non-human natural world, and in our mode of social and political life'. While environmentalism can be incorporated into other political outlooks such as socialism or conservatism, he argues, ecologism cannot because it is a distinct political outlook in its own right, and one which presents a radical challenge both to the status quo and to other political perspectives. While environmentalism is potentially compatible with modern liberal conceptions of politics and assumptions about the desirability of economic growth and progress, ecologism critiques Enlightenment ideas of human progress as hubristic and anthropocentric, and rejects all forms of industrial society (Dobson 2007: 6). From this it would follow that the movement that emerged in the early 1970s contained radical ecologist elements, while what was adopted into the mainstream was a version of environmentalism compatible with governing political and economic assumptions. However, while Dobson makes useful analytical distinctions between different political traditions, one would have to be cautious in mapping such a static typology on to historical reality, since too neat a model does not account for the messy interaction of different political perspectives at a particular historical juncture. The modern environmental movement at the turn of the 1970s both influenced, and was shaped by, the prevailing political conditions of the time.

Environmentalism's apparently simultaneous emergence in both radical protest and mainstream politics suggests that, right from the outset, it raised the question noted above regarding the relationship between elites and publics. Nixon's decision to discuss the environment in his 1970 State of the Union address is often understood as a response to 'increased public interest' in the issue (Peterson 2004: 19). In particular, the first Earth Day in 1970 is seen as having put 'pressure on the government to respond' (Vickery 2004: 119). According to one account, 'After the Earth Day demonstrations, Richard Nixon, pragmatic as always, moved not to flee or thwart but to seize upon that environmental consensus' (Wicker quoted in Vickery 2004: 119). According to another, 'After Earth Day, US President Richard Nixon…in his first State of the Union address, acknowledged a growing public awareness of environmental issues' (Haq and Paul 2012: 78). There is something curious about the chronology of such accounts, though. Earth Day was held on 22 April 1970, but Nixon's State of the Union address was delivered exactly three months *earlier*, on 22 January. How could the order of events have become confused and misremembered in this way?

There was indeed pressure on Nixon, but it came not so much from environmental campaigners as from anti-Vietnam War protestors. As Micheal Vickery (2004: 120) argues, the issue of environmental protection appealed to the president as 'an expedient means of reaching out to young, middle-class voters, thereby helping to counteract the unpopularity of the Vietnam War with that segment of the electorate'. The Earth Day organisers understood the situation in a similar way. Describing the occasion as 'channeling the energy of the anti-war protest movement', the Earth Day Network (n.d.) says the aim was to 'force environmental protection onto the national political agenda'. Similarly, Earth Day's founder, Senator Gaylord Nelson, describes the initiative as an attempt to 'infuse the student anti-war energy into the environmental cause, [so that] we could generate a demonstration that would force this issue onto the political agenda' (Nelson n.d.). For a president seen as a warmonger by large swathes of American youth, the issue did not require much forcing: it must have appeared as a godsend. Indeed, it no doubt looked that way well before the actual Earth Day protests. In November 1969 the *New York Times* was already reporting that 'Rising concern about the environmental crisis is sweeping the nation's campuses with an intensity that may be on its way to eclipsing student discontent over the war in Vietnam' (quoted in Nelson n.d.).

Nixon's attention to environmentalism, then, may be explained at least in part as ideological in much the way that, as discussed in the Introduction, Baudrillard (1974) understood at the time: an attempt to deflect from a highly divisive issue by promoting a potentially more unifying cause. Indeed, this is exactly the way that Nixon presented the issue in 1970, as 'a cause beyond party and beyond factions'; the 'common cause of all the people of this country', but particularly of 'young Americans' who would 'reap the grim consequences of our failure to act on programs which are needed now if we are to prevent disaster later' (Nixon 1970). For at least some (non-ecologist) radicals at the time, the ideological gambit was clumsily obvious. As one US civil rights activist reportedly said after Nixon's speech: 'It is a sick society that can beat and murder black people on the streets, butcher thousands in Vietnam, spend millions on arms to destroy mankind, and then come to the conclusion that pollution is America's number one problem' (quoted in Pepper 1986: 90). Yet, as we shall see, the basic theme developed by Nixon was to be of enduring importance beyond the immediate context of diverting attention from an unpopular war: in more recent years too, a key part of environmentalism's appeal for political leaders has been that it can supply a missing sense of purpose and mission. For Nixon, there had been 'too little vision' in the recent past, and the challenge was to 'inspire young Americans with a sense of excitement'. As he put it in his peroration: 'The greatest privilege an individual can have is to serve in a cause bigger than himself. We have such a cause' (Nixon 1970).

In terms of what he actually said about the environment, Nixon's 1970 address might easily be taken at first glance as an example of the sort of watered-down, compromise environmentalism that Dobson (2007) identifies as not really presenting any fundamental challenge to the established order. Alongside other, conventionally conservative, themes such as the Cold War (including the on-going war in Vietnam), and planned cuts to federal government spending but increased spending on law and order, Nixon devoted about a quarter of the speech to environmental problems. He emphasised that 'The answer is not to abandon growth' but rather to 'develop a national growth policy', since 'Continued vigorous economic growth provides us with the means to enrich life itself and to enhance our planet as a place hospitable to man'. That last pointedly anthropocentric remark is anathema to an ecological outlook, as is his characterisation of America as an 'unfinished land' which required 'perfecting' (Nixon 1970). Far from suggesting any deep green convictions, the tenor of such statements suggests a pragmatic, opportunistic use of the environmental cause. Yet this was not simply rhetoric: later that year Nixon went on to establish the Environmental Protection Agency to enforce regulation, having already set up the White House Council on Environmental Quality and the Citizens' Advisory Committee on Environmental Quality in 1969 (Lewis 1985). A closer look at the speech suggests that there was more going on.

Given that a promise of rising wealth has historically been central to the legitimacy of market societies, it is of course improbable that any leader of a capitalist democracy would stand on a platform of greater impoverishment and zero growth. Yet the rise of the modern environmentalist movement did coincide with a widespread scepticism about economic growth, including among the elite. In Nixon's 1970 speech, this scepticism was couched in terms of a concern about 'quality of life' – a phrase he repeated in different forms five times, to signal that increased material wealth was not enough. 'In the next 10 years we shall increase our wealth by 50 percent', said Nixon. 'The profound question is: Does this mean we will be 50 percent richer in a real sense, 50 percent better off, 50 percent happier?' Clearly expecting that his characterization of the present moment – 'Never has a nation seemed to have had more and enjoyed it less' – would resonate with a popular audience, Nixon explicitly framed his attention to questions of 'quality of life' as a way of connecting with people's ordinary experience: 'These are not the great questions that concern world leaders at summit conferences…[but] it is time for all of us to concern ourselves with the way real people live in real life'. His attempt to reflect people's everyday sentiments entailed acknowledging growing concern about the environment: Nixon invoked the fear that a future president might look back on the 1970s as 'a decade in which 70 percent of our people lived in metropolitan areas choked by traffic, suffocated by smog, poisoned by water, deafened by noise, and terrorized by crime'.

But Nixon's implicit scepticism about growth was also directed at a broader sense of malaise; a sense that contemporary American life lacked meaning. As the United States approached its 200th anniversary, Nixon feared that it had lost the inspiring 'spiritual quality' of the early republic: 'Today, when we are the richest and strongest nation in the world, let it not be recorded that we lack the moral and spiritual idealism which made us the hope of the world at the time of our birth'. Perhaps surprisingly, coming from a Republican president, the speech is shot through with the anti-consumerist sentiment that fulfilment does not follow from material prosperity; or in Nixon's words, with the 'recognition of the truth that wealth and happiness are not the same thing'. This was the wider context of Nixon's use of environmentalism as a potentially unifying cause: his sense that 'what this Nation needs is an example from its elected leaders in providing the spiritual and moral leadership which no programs for material progress can satisfy'. Consumer society's deficit of meaning meant that something more was needed in order to give young people 'a sense of destiny, a sense of involvement'.

These ideas were far from unusual in mainstream US politics at the time. In a 1968 speech, for example, Robert Kennedy argued that calculations of Gross National Product could measure 'the mere accumulation of material things', but not 'that which makes life worthwhile'. For Kennedy too, the concern was that the shallow pleasures of consumer society left life meaningless, so that 'even if we act to erase material poverty' that would not address the 'greater task' of confronting 'the poverty of satisfaction – purpose and dignity – that afflicts us all' (Kennedy 1968). Indeed, Nixon's predecessor in the White House, Lyndon Johnson, had voiced a similar theme in his 1964 'Great Society' speech, when he warned of America becoming 'a society where old values and new visions are buried under unbridled growth', its citizens 'condemned to a soulless wealth'.

Johnson's vision of the Great Society was of 'a place where the city of man serves not only…the demands of commerce but the desire for beauty and the hunger for community…a place where man can renew contact with nature…a place where men are more concerned with the quality of their goals than the quantity of their goods'. Linking these aspirations with the need for environmental protection, Johnson warned that:

> The water we drink, the food we eat, the very air that we breathe, are threatened with pollution. Our parks are overcrowded, our seashores overburdened. Green fields and dense forests are disappearing.…[O]nce our natural splendor is destroyed, it can never be recaptured. And once man can no longer walk with beauty or wonder at nature his spirit will wither and his sustenance be wasted.
>
> (Johnson 1964)

Addressing a college audience in this speech, Johnson directed his appeal, like Nixon, to the younger generation who might ensure that 'the demands of morality, and the needs of the spirit' were met.

Nixon's political use of the environment may indeed be understood as opportunistic and pragmatic, in the sense that he was not concerned with environmental problems simply for their own sake. Yet he was surely sincere in understanding environmental concerns as part of the larger problem that he felt compelled to address: the lack of an inspiring sense of purpose that could cohere American society, and provide meaning in a way that mere consumer satisfaction could not.

At the time of Johnson's Great Society speech this discussion was still relatively hopeful and forward-looking, but the tone became increasingly dark. Perhaps the best illustration of growing pessimism is the poster that cartoonist Walt Kelly designed for Earth Day 1970, featuring his cartoon character Pogo contemplating a littered, polluted landscape and proclaiming: 'We have met the enemy and he is us'. As Finis Dunaway (2015: 67) observes in his extended analysis of this poster, the slogan 'helped popularize environmental guilt, making this emotion central to mainstream framings of the environmental cause'. A negative, individualised sense of guilt for the ill-effects of consumer society, whereby humanity itself becomes the enemy, suggests that the rise of environmentalism was more a symptom of the felt lack of meaning and purpose in American society rather than a solution to this problem. Reputedly the 'most quoted sentence in the history of the American comic strip' (Dunaway 2015: 67), the slogan was very widely taken up. It was used in a 1969–71 exhibition, *Can Man Survive?*, at the American Museum of Natural History in New York, which linked it to the theory of overpopulation, for example; it was reportedly quoted at a meeting of Nixon's Council on Environmental Quality; and it featured in *Newsweek*'s January 1970 special issue on 'The Ravaged Environment', which argued that:

> the main villain of the piece is not some profit-hungry industrialist who can be fined into submission, nor some lax public official who can be replaced.... The villains are consumers who demand...new, more, faster, bigger, cheaper playthings without counting the cost in a dirtier, smellier sicklier world.
>
> (quoted in Dunaway 2015: 70)

Dunaway's (2015: 77) assessment is that more radical, politicised versions of environmentalism – which understood the problem in terms of the socio-economic power structures of capitalist society – tended to be eclipsed by a version which 'channelled environmentalist energies into a depoliticized quest for individual salvation'. Kelly's Earth Day cartoon was ambiguous in this respect, he argues: to some extent it 'voiced an oppositional identity that challenged the wastefulness of consumer culture'; yet at the same time it also 'encouraged audiences...to look inward and feel a sense of guilt' (Dunaway 2015: 78).

This judgement may be too generous, however, in suggesting that a critique of consumerism was 'oppositional'. The chief intellectual inspiration for the critique of consumer society developed by the 1960s New Left was the Frankfurt School, particularly Herbert Marcuse's theory of 'false needs'. By encouraging people to 'behave and consume in accordance with the advertisements', Marcuse (1972: 15) argued, consumer capitalism kept people in a state of 'euphoria in unhappiness'. Although this argument was intended as a critique of 'false consciousness', it may be better understood as itself being an instance of the way that 'the ideas of the ruling class are in every epoch the ruling ideas' (Marx and Engels 1974: 64). Rather than seeing mainstream anti-consumerism as a diluted, depoliticised version of a more genuinely radical critique, it is equally plausible to understand 'oppositional' renderings of the argument as simply a more superficially left-leaning version of what were becoming increasingly conventional ideas. Harvard economist John Kenneth Galbraith argued in his influential 1958 book *The Affluent Society* that contemporary capitalism created a 'dependence effect' whereby the supposed demands of consumers were contrived through 'advertising and salesmanship' (Galbraith 1998: 127). In more popular literature, similar ideas were articulated by journalist

Vance Packard's trio of best-selling books – *The Hidden Persuaders* (1957), *The Status Seekers* (1959) and *The Waste Makers* (1960) – which exposed the manipulative techniques of Madison Avenue, decried the competitive culture of acquisitiveness, and denounced the 'planned obsolescence' of consumer goods (Severo 1996). By 1970, a critique of consumerism had been part of mainstream discussion for some time, but rather than this indicating the growing influence of radical ideas, with hindsight it seems more likely that radicals were influenced by consumer capitalism's own sense of malaise.

As the post-war economic boom faltered and then slid into recession, environmentalist thinking provided not so much a critique as an apology for capitalist crisis. The clearest illustration of this is *The Limits to Growth*, which argued that it was nature, rather than capitalism, that limited wealth and prosperity (Meadows *et al.* 1974). Based on computer modelling by scientists at the Massachusetts Institute of Technology, the report was commissioned by the Club of Rome, co-founded by the Italian industrialist Aurelio Peccei. It was more obviously aligned with the interests of industrialists than of critics of capitalism in suggesting that an economic slowdown was a positive development.

Over time, environmentalism became understood as a clear marker of an 'oppositional', radical political identity. Alexander Cockburn, for example, a stalwart of US left-wing journalism, said he was treated as an 'intellectual blasphemer' for daring to question what had become the orthodox position on climate change in the mid-2000s (Cockburn 2008). Yet it is not immediately obvious why the Earth Day campaigners and others would have been successful in channelling left-wing anti-war politics in the direction of environmentalism. As Dobson (2007: 172–3) points out, although many writers have tried to reconcile or combine ecological and socialist perspectives, doing so requires a significant reassessment of what socialism means.

For previous generations, socialism was unambiguously understood in terms of increasing production and wealth, breaking down the barriers that capitalism presented to social and economic progress. The radical Suffragette Sylvia Pankhurst, for example, famously defined socialism in terms of 'plenty' and 'abundance': 'We do not call for limitation of births, for penurious thrift, and self-denial...[but] for a great production that will supply all, and more than all the people can consume'. Pankhurst complained that consumption was 'artificially checked' and 'cruelly limited'. She envisaged the good society as one in which 'all effort will be concentrated on supplying the popular needs', producing 'anything and everything that the people desire', including 'adornments and amusements' rather than only satisfying basic needs (Pankhurst 1923). By the 1960s, however, the Left had increasingly come to understand popular desire for 'adornments and amusements' as a problem rather than a positive aspiration for a better life.

As David Pepper (1986: 16–17) points out, left-wing politics formed only a part of the ferment of 1960s radicalism, existing alongside the very different 'hippy' counter-culture. In later accounts, the anti-capitalism of the former often gets lumped together with the latter's rejection of industrial society and its values in favour of a romantic return to nature (for example, Haq and Paul 2012: 7); and perhaps the dividing lines were not always very clear at the time either.[2] The counter-values of the hippies were consonant with a romantic ecological perspective which sought to 'break the ties of rationality and materialism and create a world of emotions and individual spontaneity' (Erisman quoted in Pepper 1986: 90). But to the extent that the Left adopted the environmental cause this implied either an attempt to persuade greens to turn red (for example, Weston 1986: 2), or else a major revision of the Left's previous positions – a revision which no doubt looked increasingly attractive as the Left's political strength declined.

Where the Left worried that its political assumptions were being called into question by the *embourgeoisement* of the working class, Daniel Bell saw a permissive hedonism as eroding traditional values from within. In his contemporaneous analysis of the 'cultural contradictions of capitalism', Bell argued that although the new 'psychedelic' counter-cultural sensibility of the 1960s might have appeared as a rejection of mainstream norms, it was in fact only another version of the hedonistic pursuit of personal satisfaction that capitalism itself promoted (Bell 1972: 36). According to Bell, this trend had been developing since the 1920s, whereby traditional values of hard work and delayed gratification (a 'Protestant Ethic' and a 'Puritan Temper') were destabilised by capitalism's own success in creating a consumer culture. The rise of consumerism seemed to offer an objective reason for the fact that, in Bell's words, 'society is left with no transparent ethos to provide some appropriate sense of purpose, no anchorages that can provide stable meanings for people' (Bell 1976: xxi).

The foregoing account has emphasised the ways in which the rise of environmentalism in the late 1960s and early 1970s was driven by elite concerns. This is not to suggest that there was some grand Machiavellian strategy on the part of political leaders, but it is to suggest that the conventional story of modern environmentalism's emergence, whereby radical grassroots protest 'forced' the issue on to the agenda of mainstream politics, is misleading. Rather, one can see a convergence of different critiques of consumerism – including New Left worries about the 'affluent worker' who could no longer be thought of as a potential revolutionary subject (Steigerwald 1995: 127); and the elite's difficulty in offering a unifying set of values and meanings to a politically divided society – which provided an opening for ecological and environmentalist ideas. It was the pull from above rather than the push from below that was more important in making these ideas part of mainstream politics. The record of politicians' engagement with environmentalism in the past helped to set the pattern for the next key moment we examine here: the emergence of climate change as a political issue.

Climate change at the end of the Cold War

Climate change became a mainstream political issue in 1988, the year of NASA scientist James Hansen's Congressional testimony on global warming, British Prime Minister Margaret Thatcher's environment speech to the Royal Society, and the establishment of the Intergovernmental Panel on Climate Change (IPCC). At the end of the year, *Time* magazine decided to change its annual 'Man of the Year' feature to name the 'Planet of the Year': our 'Endangered Earth'.[3] Rusi Jaspal and Brigitte Nerlich (2012: 123) suggest that '1988 can…be seen as a crest of a wave of social, political and media recognition of climate change as a global problem in need of global solutions', though in the British media in 1988 the 'wave' of policy and media interest was still building, toward the crescendo of the 1992 Rio Earth Summit (Mazur 1998: 467).

The context is of course highly significant: again it was a moment of great political upheaval, but of a very different nature from that of the late 1960s and early 1970s. Whereas in that earlier period the problems faced by leaders involved intense political division and contestation, in the late 1980s and early 1990s the problem was more or less the opposite: the exhaustion of Left/Right politics and the decline of public political engagement. The fall of the Berlin Wall in 1989 and the collapse of the Soviet Union in 1991 are the great symbolic events of this his-toric change, but even before the end of the Cold War it was clear to some observ-ers that the age of Left and Right, the era of modern politics which began with the French Revolution, was coming to a close. As early as 1986, Cornelius Castoriadis described the 'disappearance of social and political conflict and struggle':

> [N]ow nobody dares express a grandiose or even moderately reasonable pro-ject which goes beyond the budget or the next elections….Nobody partici-pates in a *public* time horizon….And it is not difficult to understand, I think, because people feel, and rightly so, that political ideas which are in the politi-cal market as it exists now are not worth fighting for.
>
> (Castoriadis and Lasch 1986: 20–1)

This was the context which both enabled the emergence of climate change as a mainstream political issue, and also shaped how it was articulated and used by political elites.

Margaret Thatcher is widely acknowledged as one of the first Western leaders to seek to put climate change on the political agenda (Dee 2013, Hulme 2013, Parnell 2012), despite reservations about the extent to which she was willing to match her public speeches on the issue with action (Bell 2013, Monbiot 2005). Her first and most famous intervention was her speech to the Royal Society in September 1988, although further speeches on the issue in 1989 and 1990 are more revealing in terms of the political meaning of climate change at the end of the Cold War.

In her Royal Society speech, Thatcher presented the goal of protecting the 'balance of nature' as 'one of the great challenges of the late Twentieth Century'. She nevertheless evinced confidence that the problem could be tackled ('we have shown our capacity to act effectively'), and implied that Britain was already leading the way through scientific research. Noting three atmospheric problems caused by human activity – a rise in greenhouse gases, ozone depletion and acid deposits – Thatcher warned that 'we have unwittingly begun a massive experiment with the system of this planet itself'. Her estimation of the problems was couched in terms of a pragmatic cost-benefit analysis ('Even though...action may cost a lot, I believe it to be money well and necessarily spent'). Acting to tackle environmental problems was not seen as contradicting the imperative of capitalist economic development, because 'the health of the economy and the health of our environment are totally dependent upon each other'; the conclusion also reached in the Bruntland Report the previous year (World Commission on Environment and Development 1987). Declaring that 'The Government espouses the concept of sustainable economic development', her view was that 'Stable prosperity can be achieved throughout the world provided the environment is nurtured and safeguarded' (Thatcher 1988). Although this speech is often highlighted as a key moment in the emergence of climate change as a political issue, Thatcher's presentation marked no significant departure from the sort of managerial environmentalism that Dobson (2007: 2–3) sees as readily incorporated into mainstream political discourse.

The following November, Thatcher gave a major speech on the environment to the UN General Assembly (Thatcher 1989). The problems she highlighted were similar – CO_2 emissions, the ozone layer, and also tropical deforestation – and her simultaneous advocacy of environmental protection and economic development was still there too.[4] There were, however, some important shifts in the way that she presented both the problem and potential responses to it in her 1989 speech: first, she painted a darker picture of humanity's relationship with nature; and second, she presented the challenge of climate change as the equivalent of, or a replacement for, the political and military challenges of the Cold War era.

The environmental challenge, Thatcher said, was one that 'We should embark on...hopeful of success, not fearful of failure', but at the same time she also presented that challenge in stark terms which suggested a less confident view of human progress. Like her Royal Society speech, this one still emphasised the importance of (British) scientific knowledge, but she now also called for 'humility'. Describing the nineteenth century as 'a time when men and women felt growing confidence that we could not only understand the natural world but we could master it, too', Thatcher argued that 'Today, we have learned rather more humility and respect for the balance of nature'. Human reason and science now had to be used 'to find a way in which we can live with nature, and not dominate nature'; to 'teach us today that we are not, that we must not try to be, the lords of all we survey'. Rather, human beings were 'the Lord's creatures, the trustees of this planet, charged today with preserving life itself...with all its mystery and all its wonder'. The religious undertone was echoed in Thatcher's suggestion that it was environmental change which 'drove man out of the Garden of Eden': flawed, fallen humanity itself was the problem. 'It is mankind and his activities which are changing the environment of our planet in damaging and dangerous ways', she argued; 'the main threat to our environment is more and more people, and their activities' (Thatcher 1989).

This theme was drawn out further in two speeches she delivered in 1990. Repeatedly emphasising the problem of 'very high population growth' (Thatcher 1990a) or 'the population explosion' (1990b), Thatcher again used religious phraseology ('give praise and thanks for the miracles of creation' (1990a)), and suggested that it was necessary to rethink the goal of human progress:

> For two centuries, since the Age of the Enlightenment, we assumed that whatever the advance of science, whatever the economic development, whatever the increase in human numbers, the world would go on much the same. That was progress. And that was what we wanted. Now we know that this is no longer true. We have become more and more aware of the growing imbalance between our species and other species, between population and resources, between humankind and the natural order of which we are part.... We must remember our duty to Nature before it is too late.
>
> (Thatcher 1990b)

Thatcher also argued in this speech that there was 'already a clear case for precautionary action at an international level'. These ideas about rebalancing humanity's relationship with the earth, owing a duty to nature, questioning Enlightenment assumptions about progress, and acting in accordance with the precautionary principle are characteristic of genuinely ecological views (as opposed to merely environmentalist ones).

Thatcher's adoption of these ecological themes is a striking departure in mainstream political discourse in comparison with the examples we have looked at earlier. It also illustrates very clearly the potential 'congruence' between ecologism and conservatism identified by Dobson (2007: 160–61) and by James Connelly et al. (2012: 62). For Dobson, there are nevertheless significant differences: conservatism's veneration of tradition suggests both an enduring anthropocentrism (prioritising human cultural history rather than natural history), and an emphasis on connections with the past rather than on responsibility to future generations, which distinguish it from ecological thought (Dobson 2007: 162–63). These supposedly pronounced differences, though, were not particularly apparent here. Thatcher described humanity as part of 'the natural order' and said that the 'duty to Nature will weigh on our shoulders for as long as we wish to dwell on a living and thriving planet, and hand it on to our children and theirs'. She also observed that it would take 'until the old age of my grandson, to repair the damage to the ozone layer above the Antarctic' (1990b).

Thatcher's 1989 UN speech was also distinctive in attempting to connect this conservative/ecological emphasis on human limits with a different set of problems arising from the post-Cold War unfreezing of international affairs. The Cold War was a cause with which, along with US President Ronald Reagan, Thatcher had been very closely associated over the previous decade, but it may not be obvious why she would also use it as a way to frame the problem of the environment. The point was that although the recent thaw in relations with the Soviet Union had 'brought the promise of a new dawn, of new hope' to international affairs, just at the moment when 'conventional, political dangers...appear to be receding', another danger presented itself: 'the prospect of irretrievable damage to the atmosphere, to the oceans, to earth itself' (Thatcher 1989).

The second half of Thatcher's UN speech was therefore largely devoted to elaborating a new framework of international cooperation, one designed to counter environmental threats rather than the threat of communism. The 'world community' or the 'International Community' had to come together, she urged, because 'the environmental challenge which confronts the whole world demands an equivalent response from the whole world'. Setting out a number of steps which the British government was already taking, Thatcher proposed that the UN should 'prolong the role of the Inter-governmental Panel on Climate Change' in order that it could underpin necessary agreements on international protocols and regulations to combat the problem. The negotiations to reach these agreements, she said, presented a challenge 'as great as for any disarmament treaty' and would entail 'a vast international, co-operative effort' (Thatcher 1989). Speaking in the month that the Berlin Wall came down, Thatcher adopted the one-world perspective associated with environmentalism (remarking that photographing the earth from space gave rise to 'the recognition of our shared inheritance on this planet'), but adapted it to fit the political imperative of the moment: the need to find a new basis for international cooperation, and a new international leadership role for Britain, after the collapse of the Cold War framework.

This context helps to explain the emotional tone of her two 1990 speeches, poised between fear (the felt need to invoke a new threat, comparable to or greater than nuclear war with the Soviets), and a more optimistic sense of hope for the future (the need to suggest that Britain and the West had a new and inspiring mission). This was best encapsulated in a line that was repeated almost word for word in both speeches: 'Our ability to come together to stop or limit damage to the world's environment will be the greatest test of how far we can act as a world community'.

Articulating the nature of this 'test' entailed, first, emphasising the need for international cooperation to be achieved through a reinvigorated UN: 'As East/West confrontation diminishes, as problems which have long dominated the United Nations' agenda…are being resolved, we have an opportunity to rediscover the determination that attended the founding of the United Nations' (Thatcher 1990a). The hope that, with the Cold War over, the UN would now be able to act more effectively was a very common one at the time. Since Britain is a permanent member of the Security Council, this also implied a leading role for the UK in addressing 'issues which call for a much higher level of international cooperation, more intensive than anything we have achieved so far' (1990a).

Second, as her focus on the UN implied, the necessary cooperation on the environment was understood as comparable to military cooperation, underlining the sense of an historic mission to confront a new enemy. 'Spending on the environment is like spending on defence', said Thatcher (1990a), 'if you do not do it in time, it may be too late'. Similarly, she maintained that 'the threat to our world comes not only from tyrants and their tanks', but also from the 'danger of global warming…as yet unseen, but real enough' (1990b). In referring to 'tyrants and their tanks' Thatcher was thinking, not of the USSR, but of Saddam Hussein's Iraq, which the UK and US were then poised to attack in the 1990–91 Persian Gulf

War. She spelled out the parallel by observing that the 'vital' challenge of 'defending our liberty and keeping the peace', which had defined the Cold War era, had left a legacy of successful international cooperation which could be 'seen in the Gulf, where the nations of the world have shown unprecedented unity in condemning Iraq's invasion [of Kuwait in August 1990] and taking the measures necessary to reverse it' (Thatcher 1990b).

The notion that tackling the danger of climate change was comparable to fighting the Gulf War may seem incongruous; perhaps just a peculiarity of Thatcher's militaristic, 'Iron Lady' persona. Yet a somewhat similar parallel had been drawn a decade previously by President Jimmy Carter in a 1979 televised address, in which, amid repeated allusions to American victory in World War II, he declared his determination to 'win the energy war'. The US was then in the grip of an energy crisis precipitated by a sharp rise in oil prices, so rather than addressing environmental issues more broadly, Carter's very particular message was that Americans had a national duty to conserve energy ('Every act of energy conservation…is an act of patriotism'). Indeed, although he promised to 'protect our environment', he also added bluntly that 'when this nation critically needs a refinery or a pipeline, we will build it' (Carter 1979).

The specific circumstances of 'intolerable dependence on foreign oil' threatening 'the very security of our nation' seemed to prompt Carter's war analogy in 1979, just as the then on-going process of assembling the Gulf War coalition perhaps seemed an obvious point of comparison for the sort of international cooperation on global warming that Thatcher sought to promote in 1990. But there was also another reason that Carter, rather uncharacteristically, chose to frame the issue as a 'war', 'battle', and 'fight'. In a memo to the President shortly before he was due to give the speech, his Chief of Staff, Hamilton Jordan, urged Carter to go ahead with it despite the doubts expressed by some of his advisers, because 'The energy problem symbolizes all of the problems that face our society'. Jordan characterised America as 'increasingly pluralistic, politically fragmented and dispirited about its future', and as being in a state of 'psychological panic', advising that 'the energy problem is the perfect "hook" on which to make the other points about the crisis of spirit which face [sic] our people'.[5] This is indeed exactly what Carter did, describing a 'crisis of the American spirit' in familiar terms: 'piling up material goods cannot fill the emptiness of lives which have no confidence or purpose'.

Carter's gloomy portrayal of US society, and his frank admission of 'paralysis and stagnation and drift' in Washington led to this becoming known as his 'malaise speech'.[6] The interesting thing for our discussion, though, is that the immediate and pragmatic goals of promoting 'energy conservation' and developing alternative fuel sources were presented by Carter in almost heroic terms, as the solution to a much larger political problem:

> Energy will be the immediate test of our ability to unite this nation, and it can also be the standard around which we rally. On the battlefield of energy we can win for our nation a new confidence, and we can seize control again

of our common destiny….[T]he solution of our energy crisis can also help us to conquer the crisis of the spirit in our country. It can rekindle our sense of unity, our confidence in the future, and give our nation and all of us individually a new sense of purpose.

(Carter 1979)

Thatcher's understanding of environmental damage as a 'test' for global cooperation can be understood in a similar way, as offering a new rationale for international action and a potential sense of mission or source of meaning.

As we have seen, the difficulty of giving meaning to modern life has been a recurring theme in the discussion of environmental problems by mainstream politicians. Adopting the environmentalist cause has repeatedly seemed an attractive option when other political ideas have been seen as inadequate, weak or divisive, offering a way to respond to the apparent emptiness of superficial consumerism. The decline and then sudden collapse of the Left/Right framework at the end of the 1980s re-posed this problem in a new and urgent way, as the moment of Cold War victory was undercut by an uneasy sense of terminus. Francis Fukuyama (1992: 312) voiced this ambivalence very clearly at the time, combining a triumphalist announcement of the 'end of History', with a pessimistic meditation on the life of the Nietzschean 'last man', for whom 'physical security and material plenty' could not fill the void of meaning. Liberal capitalism's historic victory simply left people wondering 'Is this really what the human story has been "all about" these past few millennia?' (Fukuyama 1992: 312).

As Zaki Laïdi (1998) observes, the problem of a post-Cold War 'world without meaning' was particularly sharply felt in terms of international relations. In a 1992 speech, for example, former President Ronald Reagan noted that the end of the Cold War had 'robbed much of the West of its uplifting, common purpose'. Speaking on the day that his successor and former Vice-President, George Bush, was sending the US marines on a mission to Somalia, Reagan's proposed solution to this loss of unifying purpose was to find a new 'cause' in humanitarian military intervention, whereby 'the world's democracies…[would] enforce stricter humanitarian standards of international conduct' (Reagan 1992). This reimagining of the West as, in Reagan's words, 'a humanitarian velvet glove backed by a steel fist of military force' proved to be a long-running theme of Western foreign policy throughout the 1990s and beyond, providing at least a temporary solution to the crisis of meaning. Thatcher's understanding of environmental damage as a 'test' for global cooperation can be understood in a similar way: as offering a new rationale for international action, and a source of meaning for domestic politics. Strikingly, the last president of the Soviet Union, Mikhail Gorbachev, made a similar move, coming up with the idea for Green Cross International in 1990.[7]

The issue was also framed in a similar way around the same time by Al Gore. Accepting the US Vice-Presidential nomination in 1992, Gore highlighted a crisis of values and meaning in American politics: 'millions of people are losing faith in the very idea of democracy and are even in danger of losing heart because they fear their lives may no longer have any deeper meaning or purpose' (Gore 1992). Given that this was an electioneering speech, it is not surprising that Gore blamed the incumbent Bush administration for letting the country down on many issues, including having 'embarrassed our nation when the whole world was asking for American leadership in confronting the environmental crisis'. The larger context, though, was clearly the new post-Cold War landscape. Reflecting that people elsewhere, 'inspired by the eternal promise of America', had recently 'torn down the Berlin Wall, brought communism to its knees and forced a racist government in South Africa to turn away from apartheid', Gore contrasted this with the situation at home, where 'we face our own crisis of the spirit here and now in America'. Like Reagan, Gore mentioned what was to become the more conventional policy response, arguing that the US had to intervene to deal with a 're-emergence of ancient ethnic and racial hatreds' in the wake of the Cold War, but the main issue through which he offered a solution to the malaise he described was of course the environment: 'The task of saving the earth's environment must and will become the central organizing principle of the post Cold War world'.

For Gore, the key idea which made the environment seem plausible as a new 'central organizing principle' was that of people being 'connected':

> [J]ust as the false assumption that we are not connected to the earth has led to the ecological crisis, so the equally false assumption that we are not connected to each other has led to our social crisis. Even worse, the evil and mistaken assumption that we have no connection to those generations preceding us or those who will follow us has led to the crisis of values we face today.
>
> (Gore 1992)

The speech was about the need for hope and renewal: 'The time has come for all Americans to be a part of the healing. In the words of the Bible, "do not lose heart." This nation will be renewed'. Yet, despite his comments about connections to past generations, for Gore this renewal could only be achieved by breaking with the past and its mistaken assumptions: 'For generations we have believed that we could abuse the earth because we were somehow not really connected to it'. Instead, the main focus was on future generations, as a means to resolve the crisis of meaning: to 'recapture...faith in the future' and to 'unite our country behind a higher calling'.

From the perspective of a Conservative politician such as Thatcher or a Democrat like Gore, then, the challenge of tackling environmental problems could be seen as a 'test' for international cooperation, an 'organising principle' for international action, a potential new source of meaning and purpose for the emerging new post-Cold War order. This is not to suggest that either Gore or Thatcher were 'using' environmentalism in an instrumental or cynical fashion, or that they did not sincerely believe their arguments. Rather, it is to suggest that for political leaders part of the attraction of environmentalism at the end of the Cold War was that it appeared to offer a promising solution to the crisis of meaning, and that this influenced how they understood and articulated the issue.

Getting emotional

In his 1992 speech, Gore made two moves which are of particular significance for this discussion: first, he explicitly framed his appeal to the electorate as non-political, declaring that 'this election isn't about politics'; second, he then talked at length about his family and his personal emotional experiences. These rhetorical moves reveal much about the characteristics of post-political communication, and about how it addresses the issue of climate change.

Gore made this speech on the day that Texan billionaire Ross Perot announced he was quitting the presidential race, and it included a 'special plea' to Perot supporters, who had 'already changed politics in this country for the better', to 'stay involved'. Perot's surprisingly successful campaign was an early example of the way that individuals, single-issue campaigns or new populist parties can upset the balance of mainstream politics in an era when established parties have only weak ties to wider society. In the years since this speech it has become increasingly common for mainstream politicians to try to adopt some kind of 'outsider' status in order to overcome public disillusionment with the political class. In Gore's case, since he was attempting to address the lack of wider sense of purpose and meaning associated with the collapse of the old Left/Right political framework, it made sense to place himself above or outside politics as conventionally understood, and his involvement with environmental campaigning perhaps made this easier since the issue does not neatly align with old party-political divisions. Instead of being about politics, according to Gore the election was about 'the responsibilities that we owe one another, the responsibilities we owe our children'. He thereby shifted the ground, from politics to the sphere of personal responsibility and ethical conduct, with the notion of care for future generations encompassing both familial ties and environmental concern.

In the section of the speech which immediately followed this non-political appeal, Gore began talking about his background and personal life. This too has become an increasingly common feature of political speeches in recent years, as leaders feel compelled, given the diminishing appeal of conventional political ideas, to present a strong personal narrative which conveys their character, values and personality. He began this part of the speech quite conventionally, mentioning his pride in his wife, parents and children, but took an unusual turn when he mentioned his experience of loss: 'I know what it's like to lose a sister and almost lose a son'. Gore devoted more than a quarter of the speech to his personal narrative, and most of it concerned his son Albert's near-fatal road accident three years previously. The heightened emotional tone and compelling detail of this passage stand out: it is the most personal, confessional moment of his own story, and, it transpires, is also the reason he can transcend the stale politics of the past and rediscover a greater sense of purpose:

> I want to tell you this straight from my heart: That experience changed me forever. When you've seen your six-year-old son fighting for his life, you

realize that some things matter a lot more than winning. You lose patience with the lazy assumption of so many in politics that we can always just muddle through. When you've seen your reflection in the empty stare of a boy waiting for his second breath of life, you realize that we were not put here on earth to look out for our needs alone. We are part of something much larger than ourselves. All of us are part of something much greater than we are capable of imagining. And, my friends, if you look up for a moment from the rush of your daily lives, you will hear the quiet voices of your country crying out for help. You will see your reflection in the weary eyes of those who are losing hope in America, and you will see that our democracy is lying there in the gutter waiting for us to give it a second breath of life.

(Gore 1992)

In this story, the country itself becomes the injured child lying in the gutter, and the would-be Vice-President is the saviour-parent of American democracy. He can take on this role because the near experience of tragic personal loss has allowed him to rise above the everyday and to glimpse a larger meaning. Gore's similar use of this aspect of his personal back-story elsewhere, including in his 2006 film *An Inconvenient Truth* (discussed in Chapter 4), suggests that its invocation here is more than simply electioneering technique and has a closer connection with his environmental message. In this speech, Gore's attempt to make an emotional connection with the electorate involves a claim to moral purpose – a claim based on an understanding of the pain of imminent potential loss and of the need to act protectively toward future generations.

As Dunaway (2015: 3) observes, 'a focus on children as emotional emblems of the future' has long been a recurring motif in environmental imagery. It is not difficult to see why that would be the case: as innocent victims of future-located environmental risks such as climate change, children are a powerful symbol of vulnerability. A 2006 US television advert made by the Advertising Council and the Environmental Defense Fund, for instance, depicted a man talking to camera about the dangers of global warming as a train speeds toward him in the background. Concluding that the consequences will only be felt in 30 years' time, he says 'that won't affect me' and steps off the rail tracks – only to reveal a young girl standing behind him, about to be flattened by the train (Dunaway 2015: 262).[8] Children can also be used to embody moral authority, as in the early 1960s 'Keep America Beautiful' campaign featuring the character of Susan Spotless, a little girl in a white dress who admonishes her parents for littering (Dunaway 2015: 84). It is notable that a month before Gore made the speech discussed above, world leaders at the UN Rio Earth Summit were addressed by a child: 12-year-old Severn Suzuki. Speaking on behalf of the Environmental Children's Organisation, which she had founded aged nine, Suzuki told the delegates she had come 'to speak for all generations to come'.[9]

The potent combination of moral authority and vulnerability has made children an irresistible symbol for political leaders, which is no doubt why the UK government's 2009 climate change adverts – highlighted at the start of this chapter – focussed entirely on children. The campaign included a series of print ads based on nursery rhymes (two of which were banned by the Advertising Standards Authority for making claims not supported by scientific evidence), with the strapline 'It's our children who'll really pay the price'; and further adverts on the theme 'Save our climate for our children', which showed aspects of children's play (building a snow-man, sailing a toy boat) that would be threatened by climate change.[10] The accom-panying television advertisement showed a father reading his daughter a bedtime storybook about the catastrophic effects of climate change, featuring illustrations of a weeping rabbit, a drowning dog and a menacing carbon monster. When the girl asks 'Is there a happy ending?' a voiceover says: 'It's up to us how the story ends'. The ad prompted hundreds of complaints from viewers, and *Nature*'s website described it as the 'Worst. Climate. Campaign. Ever.' (Cressey 2009). Defending the government's campaign, UK Department of Energy and Climate Change minister Joan Ruddock said that while the advert was 'directed at adults', the 'proposition to "protect the next generation" is a motivating one' (Sweney 2009).

Yet while there is a long history of children being used as symbols in political messages, advertising and other contexts (Holland 1992), there is also something distinctive going on in the mobilisation of child imagery in relation to climate change. Gore's implied parallel between the parent/child and the politician/voter relationships has been mirrored by social scientists seeking to explain the nature of contemporary governance in Western democracies. In his 1999 Reith Lectures series, for example, Anthony Giddens drew similar comparisons between a demo-cratic mode of everyday family life and the conduct of public, political democ-racy, characterising both in terms of a 'democracy of the emotions' (Giddens 1999, 2002). Similarly, in his work on 'emotional governance' Barry Richards (2007: 94) analyses state authority in terms of 'the power once exercised over us by parental figures'. After the collapse of the traditional ideas-driven framework of Left and Right, political leaders have sought to connect with electorates on a more thera-peutic, personal and emotional level; and, since it is now much more difficult to derive authority from traditional forms of public, political leadership, they have sometimes tended to cast themselves as quasi-parental figures with moral and emo-tional authority. In this context, children are not only significant as powerful emo-tional symbols, but also provide a kind of stand-in political constituency, whereby an emotional appeal on behalf of children and future generations can substitute for a political appeal to self-determining adults.

Although the UK government's bedtime story advert attracted an usually high volume of complaints, it was not an isolated example: the official film made to publicise the 2009 UN Climate Change Conference in Copenhagen was strikingly similar. Produced by the Danish Foreign Ministry and screened at the opening of the summit, the video literally depicts a nightmare scenario of environmental apoc-alypse: it features a young girl having a bad dream about the cataclysmic destruction of an apparently depopulated earth in a rapid series of natural disasters – drought and desertification, earthquake, tornado and flooding. Her nightmare is induced by watching television news stories about the impacts of climate change before she goes to bed. After she wakes up screaming in terror, she relates the dream to her father, who comforts her by going online and showing her the official Copenhagen Summit website, which offers reassurance that international leaders are aware of the problem. Inspired by the website, the girl borrows her father's camcorder and runs up to the roof of their apartment building, where she films herself saying 'please help the world'. The phrase, which is also the title of the film, is then reiterated by other children from around the world, their successive images zooming away to form the summit's globe logo, accompanied by the slogan 'We have the power to save the world. Now'. [11] The video provides a striking insight into how elites view themselves in relation to the issue of climate change: what appears, at first glance, to be a campaigning film about people putting pressure on world leaders is actually more like an elite wish-fulfilment fantasy, in which child-citizens across the globe put their faith in parent-politicians engaged in an heroic, planet-saving mission.

As we have argued, contrary to the common-sense assumption that politicians are only willing to take up environmental issues if they are put under sufficient pressure by activists and campaigners, today leaders appear to crave such 'pressure'. If it is not forthcoming, they will encourage and, if necessary, simulate it. Hence the British government's 2009 climate change adverts were branded, not as official government communications, but as part of a campaign called 'Act on CO_2', or 'Act on CO_2penhagen'. Packaged as if it were the initiative of an activist organization, the campaign urged: 'Pledge your support for an ambitious global deal here!' because 'We need your backing to help us negotiate'.[12] Behaving as if it were its own pressure group, the government urged the public to urge it to act. At the same time, it also berated its citizens for their apathy. Worrying that 'there isn't yet that feeling of urgency and drive and animation about the Copenhagen conference', for example, UK Foreign Secretary David Miliband complained that, for the people he wanted to be calling on him to act, 'the penny hasn't dropped that this climate change challenge is real and is happening now' (Devlin 2009).

Miliband was speaking at the launch in October 2009 of the Science Museum's 'Prove it!' exhibition – another ersatz campaign, this time inviting people to sign up to the statement: 'I want the government to prove they're serious about climate change by negotiating a strong, effective, fair deal at Copenhagen'. Yet according to the museum's director, the exhibition was created in direct response to a briefing from the government's Department of Energy and Climate Change, when 'we realised that public interest had flattened out and yet here we were approaching the most historic negotiations in human history' (Devlin 2009). Similarly, the government's controversial bedtime story advert was apparently prompted by official concern about a lack of public interest: Ruddock lamented that 'people don't realise that climate change is already under way and could have very severe consequences for their children's lives' (Cressey 2009).

Such is the relationship that has developed between the elite and the electorate in recent years: they urge citizens to be less apathetic and to urge them to act. This strange simulation of a political relationship arises from the way that environmental activism provides a vocabulary through which leaders can articulate a sense of purpose and meaning which is otherwise generally absent from today's narrow political discourse.

Conclusion: continuity and change

The importance of the environment as an issue in mainstream politics has waxed and waned over the years, in part no doubt because politicians have often been opportunistic, with only a shallow commitment to light-green environmental policies. The British Conservative leader David Cameron, for example, promised the 'greenest government ever' on coming to power in 2010, but by 2013 was reportedly telling aides to 'get rid of all the green crap' because environmental taxes added too much to the cost of energy bills (Randerson 2010, Mason 2013). In that sense it is logical for environmental campaigners to understand their goal as maintaining pressure on politicians, by engaging public interest – which has also waxed and waned – on issues such as climate change. Yet it is striking that at crucial moments – the birth of the modern environmental movement at the turn of the 1970s, and the emergence of political concern about climate change at the end of the Cold War – it has been the needs of political elites themselves which have driven a spike of interest rather than pressure from below.

As this suggests, we can see some clear continuities in the way that different generations of political leaders in both the UK and US have engaged with environmentalism. The concept of 'quality of life', elaborated in the early 1970s as a 'new tool for decision-makers' by Nixon's Environmental Protection Agency, for example, has found cross-party support among all the main British political parties in the twenty-first century.[13] Labour peer Professor Richard Layard became known as the government's 'happiness tsar' in the mid-2000s, advising prime minsters Tony Blair and Gordon Brown on 'wellbeing' (Jeffries 2008); the Liberal Democrats formally adopted 'quality of life' as a 'new purpose for politics' at their 2011 party conference;[14] and the Conservatives promoted a 'happiness agenda' both in opposition and in government, informed by the findings of the party's Quality of Life Policy Group overseen by Zac Goldsmith, the director of *The Ecologist* magazine (and nephew of its founder), who then went on to become a Conservative MP. Similarly, President Carter's idea of waging an 'energy war', echoed in Thatcher's comparison between climate change and military enemies such as the Soviet Union and Iraq, has been reworked several times. In 2004, for example, the British Government's chief scientific adviser, Sir David King, said that global warming was a 'bigger threat than terrorism', a view reiterated by President Barack Obama in 2015.[15] Indeed, climate change has for some years featured as a security threat in the National Security Strategy documents of both Britain and the United States.[16]

Both of these thematic continuities are symptomatic of the same underlying trend: the way that environmentalism has been called on time and again to resolve a 'crisis of spirit', or lack of cohesion and purpose. In his study of US presidents' environmental rhetoric, Martín Carcasson (2004: 278–80) argues that presidents should 'serve the role of moral leader and mythmaker', educating the public and developing a 'transcendent discourse' that can offer a motivational narrative or 'myth'. In the examples considered here, however – from Johnson's Great Society speech onwards – that is exactly what political leaders have been doing repeatedly, in response to a felt lack of meaning.

However, as we have argued, with the decline and collapse of Left/Right politics at the end of the Cold War, the problem of meaning was posed in a new, and much more far-reaching way. While this may have been immediately experienced in terms of the need for a new overarching framework for international cooperation, the problem ultimately ran much deeper: almost overnight, the fundamental rationale for political engagement, and the ideas through which it had been possible to articulate a vision of the future, no longer made sense. In the past, although leaders may have used environmental issues ideologically to try to cohere a divided or demoralised society, the appeal of properly ecological views was always limited while other political outlooks still clung to Enlightenment assumptions about progress. Since the late 1980s, such assumptions have all but disappeared from public political discourse. Mrs Thatcher's adoption of ecological ideas at the turn of the 1990s is one of the earliest symptoms of what has since become a more general trend: a rejection of previous conceptions of progress and a more humble appraisal of human potential.

A 1991 report by the Club of Rome expressed this shift very clearly. Noting that 'the two political ideologies which have dominated this century no longer exist', it described a 'vacuum' in which 'both the order and objectives in society are being eroded' by popular 'indifference, scepticism, or outright rejection of governments and political parties' (King and Schneider 1991: 68, 74). The report's authors observed that 'men and women need a common motivation, namely a common adversary against whom they can organize themselves and act together', and, like Thatcher, they suggested that 'pollution, the threat of global warming, water shortages, famine and the like, would fit the bill' in providing a new 'common enemy against whom we can unite'. Tellingly, however, they concluded that since such problems were really only symptoms of a deeper cause – '*human* intervention in natural processes' – the 'real enemy…is humanity itself' (1991: 75, original emphasis).

As ecologism's rejection of anthropocentrism has moved from the margins to the mainstream in the years since the end of the modern, Enlightenment era, it has become increasingly commonplace to argue that humanity itself is the problem. A 2011 symposium of Nobel Laureates, for example, addressed the issue of 'global sustainability' by putting 'humanity on trial'.[17] Although this high-level scientific meeting to produce a memorandum for the UN's Rio+20 conference was conducted in the form of a court case, with a designated judge, prosecutor and defender (and Earth as the plaintiff), the outcome of a guilty verdict was surely never in doubt.

Although there are still a few writers, such as the American sociologist Donald Gibson (2002), who critique environmentalism from a traditionally left-wing perspective, the mainstream of critical work in the humanities and social sciences now assumes the desirability of adopting a post-anthropocentric, post-human perspective as the only appropriate response to environmental crisis (see Braidotti 2013 for an exemplary overview). Giddens argued in the early 1990s that the 'Promethean outlook which so influenced Marx should be more or less abandoned in the face of the insuperable complexity of society and nature' (Giddens 1994: 79). Today, a rejection of humanism and rationalism as being too hubristic has attained the status of common sense.

For political leaders, climate change is a double-edged issue. On the one hand, there are surely limits to how far it is possible to use it as a rallying-point and source of meaning. It does provide a framework for thinking about the future but in largely negative terms, as a message of restraint and limits. The message can easily backfire when it seems too dystopian, scaremongering or sanctimonious, and it offers little opportunity for political parties to appear distinctive, given the broad consensus on the issue: in the run-up to the 2015 UK general election, for example, all three of Britain's main political parties signed a joint pledge to fight climate change regardless of who won (Harrabin 2015). On the other hand, though, climate change continues to be an attractive issue insofar as it meets politicians' need to find new, post-political ways to connect with the public and to establish a claim to leadership and authority.

In political science, the increased importance of emotions in politics over recent years is widely acknowledged (Staiger *et al.* 2010, Thompson and Hoggett 2012), and the issue of climate change is perhaps particularly well-suited to the development of a more emotional and moralistic political discourse. As we have seen, it allows leaders to envisage themselves in a quasi-parental role, protecting children and future generations, and implicitly appealing to these generations-to-come as an imagined constituency. As a follow-up to Severn Suzuki's show-stopping speech to the 1992 Rio Earth summit, for the Rio+20 meeting in 2012 there was a competition for young people to 'upload video speeches telling the world about the "Future We Want"'. The Date with History contest, described as 'an opportunity for young people from around the globe to inspire leaders and decision-makers to act boldly and urgently', was won by 17-year-old Brittany Trilford from New Zealand, who implored the assembled delegates: 'Please, lead. I want leaders who lead'.[18] The notion that young people should be called on to 'inspire' leaders to lead by pleading with them to do so implies a curious sort of political relationship, but one that politicians may well find it flattering to imagine.

Notes

1   See https://peoplesclimate.org/faqs/.
2   A study of the German Left in this era describes it as being 'torn between "hashish and revolution"' (Markovitis and Gorski 1993: 58).
3   See http://time.com/3614128/person-of-the-year-covers/.
4   '[F]irst we must have continued economic growth in order to generate the wealth required to pay for the protection of the environment…second, we must resist the simplistic tendency to blame modern multinational industry for the damage which is being done to the environment…[since] it is on them that we rely to do the research and find the solutions' (Thatcher 1989).
5   Jordan's 3 July 1979 memo is available at www.jimmycarterlibrary.gov/digital_library/cos/142099/37/cos_142099_37_17-Speech_Presidents_7-15-79.pdf.

6 See www.pbs.org/wgbh/americanexperience/features/general-article/carter-crisis-speech/.

7 True to his word, Gorbachev founded the organisation two years later. See www.gcint.org/who-we-are/our-history/.

8 Seewww.nbcnews.com/id/11992630/ns/us_news-environment/t/global-warming-ad-makes-stark-point/#.VPxA20Ks3ww.

9 The speech is available onYouTube at www.youtube.com/watch?v=oJJGuIZVfLM, with a transcript available at www.permacultureproject.com/from-the-rio-summit-a-speech-by-twelve-year-old-severn-suzuki/.

10 The banned print adverts were: 'Rub a dub dub three men in a tub, a necessary course of action due to flash flooding caused by climate change'; and 'Jack and Jill went up the hill to fetch a pail of water. There was none, as extreme weather due to climate change had caused a drought'. The ads can be viewed at http://theinspirationroom.com/daily/2010/act-on-co2-fairy-tales-2/.

11 The film is available at www.youtube.com/watch?v=NVGGgncVq-4&list=TLG 6J8w1OzZfg.

12 The site is archived at http://webarchive.nationalarchives.gov.uk/20100113202807/http://actoncopenhagen.decc.gov.uk/en/.

13 See *The Quality of Life Concept: A Potential New Tool for Decision Makers* (Environmental Protection Agency, 1973), available at http://nepis.epa.gov/EPA/html/Pubs/pubalpha_Q.html (document number 000R73002).

14 Liberal Democrats, *Quality of Life: A New Purpose for Politics* (Policy Paper 102), July 2011, http://d3n8a8pro7vhmx.cloudfront.net/libdems/pages/2008/attachments/original/1390839600/102_-_Quality_of_Life.pdf?1390839600.

15 Global warming 'biggest threat', *BBC News*, 9 January 2004, http://news.bbc.co.uk/1/hi/sci/tech/3381425.stm; Climate change a greater threat than terrorism, White House says, *ABC News*, 10 February 2015, http://abcnews.go.com/Politics/climate-change-greater-threat-terrorism-white-house/story?id=28872778. In 2014 the US Department of Defense was reportedly 'preparing for societal collapse due to climate change': *Daily Kos*, 11 August 2014, www.dailykos.com/story/2014/08/11/1320245/-The-Pentagon-is-preparing-for-societal-collapse-due-to-Climate-Change.

16 See the 2007 UK and 2010 US National Security Strategy documents, available respectively at www.gov.uk/government/uploads/system/uploads/attachment_data/file/228539/7291.pdf and www.whitehouse.gov/sites/default/files/rss_viewer/national_security_strategy.pdf.

17 Humanity on trial in Nobel Laureate court case, *Stockholm Resilience Centre*, 17 May 2011, www.stockholmresilience.org/research/research-news/2011-05-17-humanity-on-trial-in-nobel-laureate-court-case.html. The press release from the trial is available at www.sei-international.org/mediamanager/documents/News/Press-releases/SEI-PressRelease-StockholmMemorandum-18May2011.pdf.

18 See http://web.archive.org/web/20120710023556/http://datewithhistory.com:80/en/about/. Trilford's speech is available at https://www.youtube.com/watch?v=karQQb-B8Uk.

References

Baudrillard, Jean (1974) The Environmental Witch-Hunt. Statement by the French Group 1970, in Reyner Banham (ed.) *The Aspen Papers: Twenty Years of Design Theory from the International Design Conference in Aspen*. New York, NY: Praeger [available at www.metamute.org/editorial/articles/environmental-witch-hunt].

Bell, Alice (2013) Margaret Thatcher, science advice and climate change, *Guardian*, 9 April, www.theguardian.com/science/political-science/2013/apr/09/margaret-thatcher-science-advice-climate-change.

Bell, Daniel (1972) The cultural contradictions of capitalism, *Journal of Aesthetic Education, 6* (1–2): 11–38.

Bell, Daniel (1976) *The Coming of Post-Industrial Society*. New York, NY: Basic Books.

Braidotti, Rosi (2013) *The Posthuman*. Cambridge: Polity.

Carcasson, Martín (2004) Global Gridlock: The American Presidency and the Framing of International Environmentalism, 1988–2000, in Tarla Rai Peterson (ed.) *Green Talk in the White House*. College Station, TX: Texas A&M University Press.

Carter, Jimmy (1979) Crisis of Confidence, 15 July, http://web.archive.org/web/20140604053429/www.pbs.org/wgbh/americanexperience/features/general-article/carter-crisis-speech/.

Castoriadis, Cornelius and Christopher Lasch (1986) Beating the retreat into private life, *The Listener*, 27 March.

Cockburn, Alexander (2008) I am an intellectual blasphemer, *spiked*, 25 January, www.spiked-online.com/review_of_books/article/4357#.VOI2VkKs3wx.

Connelly, James, Graham Smith, David Benson and Clare Saunders (2012) *Politics and the Environment* (Third Edition). Abingdon: Routledge.

Cressey, Daniel (2009) Worst. climate. campaign. ever., *Nature News Blog*, 9 October, http://blogs.nature.com/news/2009/10/worst_climate_campaign_ever_1.html.

Dee, Jon (2013) How Margaret Thatcher led the way on climate change, Australian Broadcasting Corporation, 9 April, www.abc.net.au/environment/articles/2013/04/09/3732680.htm.

Devlin, Hannah (2009) Foreign Secretary David Miliband accuses public of climate change apathy, *The Times* (London), 23 October, www.thetimes.co.uk/tto/environment/article2144727.ece.

Dobson, Andrew (2007) *Green Political Thought* (Fourth Edition). Abingdon: Routledge.

Dunaway, Finis (2015) *Seeing Green: The Use and Abuse of American Environmental Images*. Chicago, IL: University of Chicago Press.

Earth Day Network (no date) Earth Day: The history of a movement, *Earth Day Network*, www.earthday.org/earth-day-history-movement.

Fukuyama, Francis (1992) *The End of History and the Last Man*. New York: Free Press.

Galbraith, John K. (1998) *The Affluent Society* (Fortieth Anniversary Edition). New York, NY: Houghton Mifflin.

Gibson, Donald (2002) *Environmentalism: Ideology and Power*. Huntington, NY: Nova Science Publishers.

Giddens, Anthony (1994) *Beyond Left and Right: The Future of Radical Politics*. Cambridge: Polity.

Giddens, Anthony (1999) Family, *BBC Reith Lectures 1999*, http://news.bbc.co.uk/hi/english/static/events/reith_99/week4/week4.htm.

Giddens, Anthony (2002) *Runaway World: How Globalisation Is Reshaping Our Lives* (Second Edition). London: Profile Books.

Giddens, Anthony (2009) *The Politics of Climate Change*. Cambridge: Polity.

Gore, Al (1992) 1992 VP Acceptance DNC, New York, July 16, *Speeches USA*, www.speeches-usa.com/Transcripts/al_gore-1992dnc.htm.

Haq, Gary and Alistair Paul (2012) *Environmentalism Since 1945*. Abingdon: Routledge.

Harrabin, Roger (2015) Party leaders make joint climate commitment, *BBC News*, 14 February, www.bbc.co.uk/news/science-environment-31456161.

Holland, Patricia (1992) *What Is a Child? Popular Images of Childhood*. London: Virago.

Hulme, Mike (2013) Climate change is 25 years old, *Climate News Network*, 27 September, www.climatenewsnetwork.net/tag/margaret-thatcher/.

Jaspal, Rusi and Brigitte Nerlich (2012) When climate science became climate politics: British media representations of climate change in 1988, *Public Understanding of Science*, 23 (2): 122–41.

Jeffries, Stuart (2008) Will this man make you happy?, *Guardian*, 24 June, www.theguardian. com/lifeandstyle/2008/jun/24/healthandwellbeing.schools.

Johnson, Lyndon B. (1964) The Great Society, University of Michigan, 22 May, www.pbs. org/wgbh/americanexperience/features/primary-resources/lbj-michigan/.

Kennedy, Robert F. (1968) Remarks at the University of Kansas, 18 March, www.jfklibrary. org/Research/Research-Aids/Ready-Reference/RFK-Speeches/Remarks-of-Robert-F-Kennedy-at-the-University-of-Kansas-March-18-1968.aspx.

King, Alexander and Bertrand Schneider (1991) *The First Global Revolution: A Report by the Council of the Club of Rome*. London: Simon & Schuster.

Laïdi, Zaki (1998) *A World Without Meaning*. London: Routledge.

Lewis, Jack (1985) The Birth of EPA, originally published in *EPA Journal*, November, United States Environmental Protection Agency, https://archive.epa.gov/epa/aboutepa/birth-epa.html.

Marcuse, Herbert (1972) *One-Dimensional Man*. London: Abacus.

Markovitis, Andrei S. and Philip S. Gorski (1993) *The German Left: Red, Green and Beyond*. Cambridge: Polity.

Marx, Karl and Friedrich Engels (1974) *The German Ideology*. London: Lawrence & Wishart.

Mason, Rowena (2013) David Cameron at centre of 'get rid of all the green crap' storm, *Guardian*, 21 November, www.theguardian.com/environment/2013/nov/21/david-cameron-green-crap-comments-storm.

Mazur, Allan (1998) Global environmental change in the news: 1987–90 vs. 1992–6, *International Sociology*, 13 (4): 457–72.

Meadows, Donella H., Dennis L. Meadows, Jørgen Randers and William W. Behrens III (1974) *The Limits to Growth: A Report for the Club of Rome's Project on the Predicament of Mankind*. London: Pan.

Monbiot, George (2005) Then…and now, *Guardian*, 30 June, www.theguardian.com/environment/2005/jun/30/climatechange.climatechangeenvironment1.

Nelson, Gaylord (2009) How the first Earth Day came about, *Envirolink*, 9 April, Institute of Food and Agricultural Sciences and University of Florida, https://blogs.ifas.ufl.edu/pinellasco/2009/04/09/how-the-first-earth-day-came-about/.

Nixon, Richard (1970) Annual Message to the Congress on the State of the Union, 22 January, www.presidency.ucsb.edu/ws/?pid=2921.

Pankhurst, Sylvia (1923) Socialism, *Workers' Dreadnought*, 28 July, www.marxists.org/archive/pankhurst-sylvia/1923/socialism.htm.

Parnell, John (2012) Margaret Thatcher: The green lady?, *Climate Home*, 6 January, www.climatechangenews.com/2012/01/06/margaret-thatcher-the-green-lady/.

Pepper, David (1986) *The Roots of Modern Environmentalism*. London: Routledge.

Peterson, Tarla Rai (2004) Introduction: Environmental Communication Meets Presidential Rhetoric, in Tarla Rai Peterson (ed.) *Green Talk in the White House*. College Station, TX: Texas A&M University Press.

Randerson, James (2010) Cameron: I want coalition to be the 'greenest government ever', *Guardian*, 14 May, www.theguardian.com/environment/2010/may/14/cameron-wants-greenest-government-ever.

Reagan, Ronald (1992) Better tomorrows as a noble vision approaches fruition, *Sunday Times*, 6 December.

Richards, Barry (2007) *Emotional Governance*. Basingstoke: Palgrave.

Roser-Renouf, Connie, Edward W. Maibach, Anthony Leiserowitz and Xiaoquan Zhao (2014) The genesis of climate change activism: From key beliefs to political action, *Climatic Change*, 125 (2): 163–78.

Severo, Richard (1996) Vance Packard, 82, challenger of consumerism, dies, *New York Times*, 13 December, www.nytimes.com/1996/12/13/arts/vance-packard-82-challenger-of-consumerism-dies.html.

Staiger, Janet, Ann Cvetkovich and Ann Reynolds, eds. (2010) *Political Emotions*. Abingdon: Routledge.

Steigerwald, David (1995) *The Sixties and the End of Modern America*. New York, NY: St. Martin's Press.

Sweney, Mark (2009) More than 200 complaints at government's climate change TV ad, *Guardian*, 16 October, www.theguardian.com/media/2009/oct/16/complaints-government-climate-change-ad.

Thatcher, Margaret (1988) Speech to the Royal Society, *Margaret Thatcher Foundation*, www.margaretthatcher.org/document/107346.

Thatcher, Margaret (1989) Speech to United Nations General Assembly, *Margaret Thatcher Foundation*, www.margaretthatcher.org/document/107817.

Thatcher, Margaret (1990a) Speech to the Aspen Institute (Shaping a New Global Community), *Margaret Thatcher Foundation*, www.margaretthatcher.org/document/108174.

Thatcher, Margaret (1990b) Speech at Second World Climate Conference, *Margaret Thatcher Foundation*, www.margaretthatcher.org/document/108237.

Thompson, Simon and Paul Hoggett, eds. (2012) *Politics and the Emotions: The Affective Turn in Contemporary Political Studies*. London: Continuum.

Vickery, Micheal R. (2004) Conservative Politics and the Politics of Conservation: Richard Nixon and the Environmental Protection Agency, in Tarla Rai Peterson (ed.) *Green Talk in the White House*. College Station, TX: Texas A&M University Press.

Westcott, Kathryn (2012) Are these the most offensive ads of all time?, *BBC News Magazine*, 30 May, www.bbc.co.uk/news/magazine-18243577.

Weston, Joe (1986) Introduction, in Joe Weston (ed.) *Red and Green: A New Politics of the Environment*. London: Pluto Press.

World Commission on Environment and Development (1987) *Our Common Future* (The Bruntland Report), United Nations, www.un-documents.net/our-common-future.pdf.

2

CYCLES, ARENAS AND NORMS

Understanding news coverage

For most people the news media are the most important source of information about climate change and environmental campaigning, but as well as reporting on the initiatives of others news organisations have also sometimes adopted a campaigning stance themselves. This chapter considers the news media as communicators of ideas and information, policies and programmes, claims and counter-claims about climate change which originate elsewhere; Chapter 3 will consider the media as advocates in their own right who attempt to engage and influence others. The line between these two roles is not always easy to draw, but each implies a different view of what the media's job ought to be – neutral observers and balanced reporters informing the public; or engaged and partisan campaigners, rousing public opinion and influencing policy. Moreover, it is also debateable what either of these options should look like in practice. Would neutral and balanced reporting mean reflecting all viewpoints equally, for example; or should those viewpoints be treated differently according to an estimate of their broader social resonance and level of public support; or should they be weighted according to how far they accord with the consensus of scientific opinion on the issue? Similarly, if we see the role of the media as one of engaging the public and attempting to influence debate, that raises the question of what might be the best way to do so: in the past, it has often been assumed that the priority is to build agreement around the causes of climate change and the best options for responding to it, but the post-political critique suggests that this consensus-oriented approach is insufficient or even counter-productive.

This chapter offers an overview of the way that media attention to the environment has risen and fallen over time, examining how scholars have explained such fluctuations and interrogating the different understandings of the news media that underpin these explanations. As we saw in Chapter 1, the common assumption that politicians have only been led to take up environmental issues under pressure from protestors and campaigners does not seem to fit the historical record. There is a similar assumption made in relation to the media. Indeed, one of the motivations for research on the news media and climate change is the perception that coverage has the potential to influence policy-makers, either directly or via influence on public opinion. Yet, as we shall see, the available evidence points strongly in the opposite direction, suggesting that media attention to the issue has been elite-led. Much of the research on the media and climate has either ignored, or has drawn only selectively on, critical scholarship about the news media, resulting in an incoherent understanding of the relationship between media and political power. It also rests, as some critics have pointed out, on a mistaken assumption that the end goal should be to narrow the range of 'legitimate' viewpoints available in the media (Carvalho and Peterson 2012, Pepermans and Maeseele 2014).

Explaining patterns of change

That media attention to environmental issues rises and falls over time was suggested theoretically very early on, and has subsequently been confirmed empirically by a number of studies. In his survey of longitudinal studies of news coverage of the environment, Anders Hansen notes 'two important characteristics':

> a) that once introduced in the 1960s, the 'environment' has consolidated itself as a news-beat and category of news coverage, and b) that news coverage of the environment has gone up and down in regular cycles.
>
> (Hansen 2015: 211)

The first characteristic means that even though there have been periods when the media have paid relatively little attention to the environment, it has 'never vanished completely', and historically the 'overall trend has been up' (Hansen 2015: 211). The second characteristic, however, seems to present something of a puzzle: given that 'the ups and downs in media attention to environmental issues seem to bear little relationship to the severity of these issues' (2015: 213), what leads the news media to devote great attention to the environment at some points but to neglect it at others? As Hansen goes on to observe, Anthony Downs's notion of an 'issue-attention cycle' is one of the most widely cited and influential frameworks for explaining the 'longitudinally oscillating patterns of media attention to environmental issues' (2015: 213). The attraction of Downs's model, Hansen suggests, is obvious: since studies have consistently shown that media attention to environmental problems has risen and fallen in 'cycles' over time, and since those cycles do not correlate with objective changes in the urgency of environmental problems, there must be some other logic at play – perhaps to do with the way that the media and/or public opinion work.

In his 1972 article, 'Up and down with ecology – the "issue-attention cycle"', Downs was only indirectly concerned with the news media, as one factor influencing his main object of interest, namely patterns of (American) public attention to (domestic) issues. He was prompted to take the environment as his main illustrative example by the 'remarkably widespread interest' it was attracting at the time: a change in attitudes that was occurring 'much faster than any changes in the environment itself' and which therefore seemed to demand some explanation (Downs 1972: 38). With hindsight, his article seems prescient in anticipating a waning of attention to the issue which had yet to occur. He was led to make such a prediction because his starting premise was that ebbs and flows of public interest in any given issue or problem are not driven by changes in the objective seriousness or urgency of the issue itself, but rather shift in accordance with an inner logic of public attention which applies to the progress of any issue in the public sphere. Downs postulated that initial enthusiasm for solving a problem usually assumes that to do so will not entail 'any fundamental reordering of society', and that as the realisation dawns that properly tackling it would entail significant costs and sacrifices, interest gradually starts to decline until finally the issue ends up in 'a twilight realm of lesser attention or spasmodic recurrences of interest' (1972: 39–40). In relation to the environment specifically, Downs placed it in 1972 at the mid-point where enthusiasm was beginning to be tempered by a realisation that the problem might be difficult to solve (1972: 43).

Downs identified a variety of reasons for the rise of the environment as a public issue in the early 1970s, including: people observing everyday evidence of environmental degradation such as smog, oil spills and traffic congestion; a better economic standard of living raising people's aspirations beyond problems of mundane survival; and increased media emphasis on the topic (1972: 43–5). Yet he also intimated that it was pre-eminently an elite concern, observing that 'The elite's environmental deterioration is often the common man's improved standard of living' (1972: 44). He had in mind such things as car ownership, housing development and holidays: as more people were able to enjoy these material benefits, so it was becoming 'difficult for the wealthy to flee…to places of quiet seclusion, because so many more people can afford to go with them' (1972: 44). In line with our account in Chapter 1 of how the environment emerged as a political issue in this era, Downs also noted that it was something that 'politicians can safely pursue…without fearing adverse repercussions', since it was 'not politically divisive' and was 'much safer than attacking racism or poverty' (1972: 47). Tackling the latter, he noted, 'would require millions of Americans to change their own behavior patterns, to accept higher taxes, or both', whereas in the case of the environment, technological improvements might avert the need for such far-reaching changes (1972: 47–8). As he put it:

> To the extent that pollution can be reduced through technological change, most people's basic attitudes, expectations, and behavior patterns will not have to be altered. The traumatic difficulties of achieving major institutional change could thus be escaped through the 'magic' of purely technical improvements.
>
> (Downs 1972: 48)

While this scenario may seem plausible in relation to a problem such as pollution in US cities, however, many would object that the problem of climate change is on an altogether different scale. It remains a point of contention for environmentalists whether to seek to develop some sort of 'technological fix' (or series of fixes) for climate change, or whether to encourage exactly the sort of larger social changes involving people's attitudes, expectations and behaviour which Downs suggests are so challenging that they are likely to result in declining public engagement with the issue.

Not everyone is enamoured with Downs's 'issue-attention cycle' model. Alison Anderson (2009: 169), for example, describes it as 'too linear and inflexible' and as focussing narrowly on 'the media and public agendas'. She argues that researchers need to take a wider view of the variety of actors who can potentially define 'the important issues of the moment' (2009: 169). Similarly, Max Boykoff (2009: 448) says that Downs provides 'only partial, and overly linear explanations' of a complex process. Both authors prefer the approach of Stephen Hilgartner and Charles Bosk (1988: 54), who aim to improve on crude '"natural history" models that describe the stages in the career of a social problem', such as Downs's. In fact it is probably Hilgartner and Bosk who have been more significant in the field of media and environment research. Notwithstanding the importance given to Downs by Hansen and others, his influence mainly derives from his having identified very early on what has become a long-standing research question – how to explain fluctuations in media attention to the environment – rather than from the specifics of how he himself answered it. The overall approach of much subsequent research on this question is much closer to Hilgartner and Bosk's 'public arenas' model (Lester 2010: 46–7).

Hilgartner and Bosk's model emphasises, first, that rather than individual issues unfolding according to a typical and predictable pattern, multiple issues – and the groups who define and promote them – exist in competition with one another; and second, that this dynamic process of competition takes place within and across specific institutional 'arenas' (such as newspapers, films and TV programmes, political organisations or government committees). Each arena is subject to its own conventions influencing how it prioritises issues, and all of them have a finite 'carrying capacity' in terms of the number of issues they can focus on at any one time (Hilgartner and Bosk 1988: 56). To understand the fluctuations in attention to an issue, Hilgartner and Bosk (1988: 57) suggest, one would need to examine the behind-the-scenes work of what they call 'operatives' – individuals and groups promoting particular problem definitions – and to analyse the characteristics of the particular institutional arenas within which competition for attention takes place. In terms of the latter, they suggest that four key factors influence the selection and prioritisation of issues. First, they note the importance of 'drama', though arguing that it works best when combined with claims to expertise: 'Thus, in social problem claims, officially certified "facts" are coupled with vivid emotional rhetoric' (1988: 62). Second, they remark that while the use of classic tropes is common, such as finding contemporary variations of the 'Everyman' of medieval morality plays, nevertheless 'novelty is also an important factor', and if claims become too repetitive, an over-saturation will lead to boredom and declining interest (1988: 62). Third, they observe that decisions about which issues should be prioritised are influenced by 'widely shared cultural preoccupations and political biases', not only in the sense that particular problem claims or definitions may be more culturally resonant, but also in the sense that there may be 'powerful political and economic interests' who act as sponsors for issues or actively oppose other problem definitions (1988: 64). Finally, they note the importance of organisational characteristics such as, in the case of the media, the editorial conventions of news selection, the structure of the news industry, or 'linkages between newspaper directors and other parts of the American business elite' (1988: 65).

Hilgartner and Bosk only mention the environment in passing, as one of a number of examples, but it is worth noting that they use it to illustrate that there is a potential for symbiosis and mutual reinforcement, as well as rivalry and competition, across different institutional 'arenas':

> Thus, environmental groups, industry lobbyists and public relations personnel, politicians who work on environmental issues, environmental lawyers, environmental editors, and officials in government environmental agencies all generate work for one another. At the same time, their activities collectively raise the prominence of the environment as a source of social problems.
> (Hilgartner and Bosk 1988: 68–9)

It is notable that Hilgartner and Bosk – writing, like Downs, at a time of growing public attention to the environment – seem to assume a collective elite effort to amplify the importance of environmental issues. This is more or less the opposite of why their model has seemed attractive to later researchers working on climate change coverage, however. Hilgartner and Bosk's model is subtle and flexible, able to take in a multitude of factors that affect how far issues attract and sustain public attention, including structural economic and political constraints on the media. The overall emphasis of the approach, though – at least as it has been used by others – is to suggest that the media are a relatively neutral 'arena' within which competition between different claims-makers takes place. To appreciate the significance of the approach articulated by Hilgartner and Bosk for subsequent research on the media and the environment, we can return to Anderson's article cited above.

Anderson maps out the research field in terms drawn from the discipline of Media and Cultural Studies, identifying three overlapping but distinct analytical frameworks which can help to account for patterns of media coverage of climate change: 'the political economy approach, the structuralist approach and the culturalist approach' (Anderson 2009: 170).[1] In relation to climate change, the political economy approach suggests that 'mainstream media are likely to steer clear of reporting which deeply challenges corporate interests that have a strong connection with their outlets', although Anderson says that it would be 'far too simplistic to suggest that ties with fossil fuels industries completely prevent critical reporting' (2009: 170). The structuralist approach highlights the importance of political rather than simply economic factors. It draws attention to news production routines, such as the reliance on powerful sources who act as 'primary definers' of the news agenda; and to the importance of professional identities which might suggest a 'neutral' or an 'advocacy' role for the journalist (2009: 172–3). Finally, the culturalist tradition emphasises the ways that 'cultural norms or "givens" are embedded within media coverage', for example in the use of different discursive repertoires to frame the issue of climate change in particular ways, thereby reinforcing certain meanings rather than others (2009: 174–5).

Anderson's overall argument is that we need to investigate the competitive processes of claims-making in different arenas if we are to understand 'the ebbs and flows in reporting climate change over time' (2009: 167). She maintains that 'examining the behind-the-scenes battle between news sources is crucial to understanding the pattern and nature of climate change news over time'; and that 'claims-makers are engaged in continual definitional struggles, requiring an in-depth and multi-faceted analysis of the factors influencing the effectiveness of media strategies over time'; and again that 'in-depth interviews with news sources would provide us with a greater understanding of competition to control the issue and the behind-the-scenes factors influencing patterns of reporting'; and yet again that 'there is an urgent need to uncover the "behind-the-scenes" factors that shape current representations' (Anderson 2009: 171, 173, 176, 178; and see also Anderson 2014: 49; Lester 2010: 77, 92, 180). Despite the quite different theoretical framing and the specific focus on coverage of climate change, the research agenda set out by Anderson sounds strikingly similar to that proposed twenty years previously by Hilgartner and Bosk. Can fluctuations in media coverage of the environment really be such a puzzle that we are still asking the same questions?

Indexing: the elephant in the room

What is at stake in the different approaches discussed by Anderson is how to understand the relationship between the media and power. The aim of the 'political economy approach' is to show how the spontaneous operation of the market tends to undermine the 'democratic postulate' that 'the media are independent and committed to discovering and reporting the truth, and that they do not merely reflect the world as powerful groups wish it to be perceived' (Herman and Chomsky 2002: lix). Similarly, the 'structuralist' approach suggests that the media have only a limited degree of independence ('relative autonomy'), and tend to follow the lead of powerful news sources who act as 'primary definers' of the news agenda (Hall *et al.* 1978: 57). The 'culturalist' label is a little less clear-cut because unlike the other two it is not self-adopted but imputed by others, and also because the work of so-called 'culturalist' writers tends to be more concerned with historical analysis than with theory-building, but nevertheless this approach is again broadly concerned with how 'dominant' or 'hegemonic' cultures are established (Williams 1977: 121–7). Whereas all three approaches indicate a strong elite influence on the news media, research on coverage of climate change has instead tended to assume that the elite have to be persuaded and pressured to acknowledge the issue.

In fact those ebbs and flows may not be such a great mystery: peaks in media attention to the environment have consistently followed the agenda of mainstream politics, from an 'initial peak in the early 1970s', through 'another dramatic increase in the latter half of the 1980s', to the 'considerable new resurgence in the first decade of the 2000s' (Hansen 2015: 211). Almost immediately after Nixon's 1970 State of the Union address discussed in Chapter 1, for example, *Newsweek*'s cover story was 'The Ravaged Environment' (26 January 1970); four days later the cover of *Life* magazine declared 'Ecology becomes everybody's issue' (30 January 1970); and three days after that, *Time*'s cover story was 'Fighting to save the earth from man' (2 February 1970), featuring leading environmental campaigner Barry Commoner.[2]

It might be argued that although a generally 'light-green' environmentalism was adopted in the early 1970s, that does not necessarily tell us much about responses to the problem of climate change in more recent years. Yet there is a good case that here too elite actors have played an important agenda-setting role.

Studies of the first sharp increase in media coverage of climate change in 1988 note two important factors: it was a particularly hot summer, and there were a number of statements and speeches on the issue in mainstream politics. We should not underestimate the importance of weather events as providing a short-term news hook, but political initiatives are the more significant and consistent influence on coverage. Rusi Jaspal and Brigitte Nerlich (2012: 123), for example, note that 'Key political figures from the US (President George Bush and Governor Michael Dukakis), the Soviet Union (President Mikhail Gorbachev) and the UK (Prime Minister Margaret Thatcher) all publicly attributed importance to addressing the climate change problem' in 1988, with the result that news coverage quadrupled compared with the previous year. They go on to observe that this was followed by 'a veritable avalanche in media reporting' in 1990, the year of the first IPCC report; and by very extensive coverage in 1992, the year of the Rio climate change summit (Jaspal and Nerlich 2012: 123–4). Similarly, Allan Mazur's (1998: 466) analysis of the 'flood of journalistic attention in 1987–90' notes the importance of James Hansen's 1988 congressional testimony and the fact that climate change became an election issue that year, with George Bush announcing that he would be an 'environmental president'. As Anderson (2009: 168) herself notes, 'The late 1980s saw the first sharp rise in coverage in the British and US national daily newspapers as the issues became increasingly politicized'. Other studies have shown the same thing in both a US context (Boykoff and Boykoff 2004: 130) and the UK (Carvalho and Burgess 2005: 1462–4). The issue was taken up by elite political figures and the media followed.

In any other area of coverage, this would likely be understood as an example of 'indexing', the phenomenon famously described by W. Lance Bennett (1990), whereby the news stories highlighted, and the range of opinions about them which are aired in the mainstream media, are 'indexed' to the concerns and views of elite sources. Others have provided support for this theory, especially in relation to foreign policy and international intervention, the area of coverage discussed by Bennett (see, for example, Mermin 1999). Yet when it comes to coverage of climate change, this appears to be a blind spot. Even when Bennett's work is prominently cited (for example in Boykoff and Boykoff 2007), his best-known contribution, the indexing hypothesis, is ignored.

Instead, either the media tend to be understood as a relatively neutral 'arena' or 'battlefield' in which different views compete; or else influence is seen to flow in the opposite direction, not from politicians to the media but vice versa. Jaspal and Nerlich (2012: 124) argue that 'media engagement with climate change in 1988 would have set the tone for political and public engagement for some time to come', for example; and Mazur (1998: 458) argues that 'the most widely accepted effect of news media on opinion is "agenda-setting", the placing of certain issues or problems foremost in the minds of people, including policy-makers'. While the notion that the media set the agenda in terms of what audiences are concerned with is well-established (McCombs and Shaw 1972), however, the notion that they also set the policy agenda has found little empirical support (Robinson 2002). The proposition that news coverage affects 'public worries' is very different from the idea that it also determines the agenda for 'governmental action' (Mazur 1998: 458); and it does not really make sense to treat 'policy actors' as simply part of the public (Boykoff and Boykoff 2007: 1191).

As we have seen, for Hilgartner and Bosk in the late 1980s, as for Downs in the early 1970s, attention to the environment as a public issue was implicitly understood as following elite concerns. For more recent analysts, however, the assumption tends to be that elite interests work against those seeking to promote climate change as an issue for public attention. Discussing the political economy approach, for example, Anderson (2009: 170) notes that 'Media corporate interests and the vested interests of advertisers...may discourage criticism of government's inaction over climate change or industry's role'. In fact governments have often promoted action over climate change, businesses and industries have frequently adopted a 'green' social-responsibility agenda (Heartfield 2008), and the media have, by Anderson's own account, increasingly publicised the issue over many years. Following the first peak in coverage at the end of the 1980s, there was then 'a second steep rise in coverage between 1997 and 2003', which Anderson (2009: 168) attributes to the signing of the international Kyoto Protocol on climate change in late 1997. This rise in attention was then immediately followed by a further rise as, 'between 2003 and 2006, there was a steady increase in attention among the opinion-leading newspapers in the United States, with an even more marked rise in the United Kingdom' (Anderson 2009: 169), so that broadsheet newspaper coverage increased by two and a half times between 2003 and 2006 in the US, and by four times in the UK (Boykoff and Roberts 2007: 6). Anderson (2009: 169) notes the importance of Gore's *An Inconvenient Truth* (2006) at the end of this period, and we might also mention the fact that UK prime minister Tony Blair adopted climate change as a priority in his capacity as chair of the G8 and president of the EU in 2005, and that the UK government-commissioned *Stern Review on the Economics of Climate Change* was published in 2006.[3]

Only if we assume that political elites would rather ignore climate change and are reluctant to engage with it would media attention to the issue seem puzzling.

That is not to say that such an assumption is entirely without foundation: in part what underlies this view is a concern that climate 'sceptics' or 'deniers' are funded and promoted in the media by political or fuel-industry lobbies. Yet this is not necessarily incompatible with the indexing hypothesis. Again, in other areas of news coverage the parameters of debate and disagreement over issues have been shown to shift and change in line with divisions among elite sources (Hallin 1989).

In other words, while it may well be true that 'powerful industry groups, special interest lobbies and PR companies...have manipulated scientific claims and exploited the news media' (Anderson 2009: 170–71), that does not necessarily indicate that climate change is not an elite-driven issue, only that elite opinion has been divided to a greater or lesser extent at different moments. Boykoff and Boykoff's (2007: 1196) study of climate change coverage in the US media shows a rise in 1988 when George Bush made an election campaign pledge to 'fight the greenhouse effect with the White House effect', but they observe that a further increase in coverage in 1990 was associated with the emergence of 'a coherent and cohesive group' of 'climate contrarians' who challenged reports issued that year by the IPCC, so that reporting featured 'prominent scientists and politicians from both sides of the issue'. A rise in coverage in 1992 again reflected elite dissensus – 115 heads of state gathered for the Rio Earth Summit, while President Bush and his Chief of Staff John Sununu signalled that the US was unwilling to engage with the issue (2007: 1197). Similarly, while a rise in coverage in 1997 was associated with President Bill Clinton's unexpected show of interest in 'environmental stewardship and climate change', a further jump in coverage in 2001–2 reflected the emergence of a sharp divergence between US and European leaders over the issue (2007: 1198–9); and yet another increase in coverage in 2004 repeated the pattern (2007: 1199–1200). Indeed, their study could stand as a textbook example of indexing.

The low points of coverage have been less studied than the peaks, though they have received some attention. Mazur (1998: 459) attempts to account for why there was 'a flood of journalistic attention in 1987–90 and a drought of coverage in 1992–6'. One reason, he suggests, is that after the 1992 US election there was a 'loss of news sources in Congress' as figures who had 'promoted global issues to journalists' were 'no longer available' (1998: 468). One obvious exception to this is Al Gore, who became Vice President in 1992 and therefore had a potentially stronger platform from which to promote the issue, but Mazur argues that Gore simply 'submerged his own concerns beneath those of President Clinton' (1998: 468). Similarly, Anabela Carvalho and Jacquelin Burgess (2005: 1464) note that there were 'no significant public pronouncements by political leaders' in the UK in 1991. Carvalho and Burgess suggest that 'editorial fatigue' set in as a dramatic climate catastrophe failed to materialise (2005: 1464); while Sheldon Ungar (1992: 496), despite discussing a range of political factors, ultimately concludes that the key change was that 'the weather abated' after the hot summer of 1988. Mazur (1998: 469–70) ends up declaring that 'we do not understand' the decline in coverage after 1991, and concludes that 'The media are fickle'.

It is possible that methodological differences lead Mazur to see a 'drought' of coverage in the US in 1992 while in Boykoff and Boykoff's (2007: 1197) account 'coverage continued to increase substantially' that year, but there certainly was a change: as we have noted, news coverage in 1992 was concerned with political divisions over the issue rather than with elites continuing to promote it as they had done in the recent past. If we recall the context in which the issue was initially taken up by political leaders after the end of the Cold War, however, this shift may again be understood in terms of indexing. As discussed in Chapter 1, political leaders in the late 1980s and early 1990s spoke of climate change as a challenge that was comparable to the East/West conflict of the Cold War era, as offering a new sense of historic mission and a new framework for international cooperation. Yet very soon after this, other replacements for the Cold War presented themselves in a more obvious and urgent way: war with Iraq in 1991; diplomatic, humanitarian and ultimately military intervention in the former Yugoslavia from 1991–95; humanitarian/military interventions in Somalia in 1992, and in Haiti in 1994. The significance of these events for the issue of climate change is not so much that there was a straightforward competition for attention in the media and other arenas with limited 'carrying capacity', but more that they answered the specific political needs of post-Cold War Western leaders in a clearer and more immediate way (see further Hammond 2007).

Questions of quality

Mazur (1998: 458–9) argues that the volume of coverage of climate change is more important than its quality, maintaining that 'Precisely what is said in the stories matters relatively little compared to the simple quantity of exposure'. His view gets some support from Jaspal and Nerlich (2012: 137), who suggest that while media coverage of climate change is a 'representational field' which has always been characterised by 'incoherence, tension and ambivalence', there is nevertheless 'a consensual "reality" in relation to climate change, namely that it exists as a socio-environmental issue to be debated and discussed and that one must take a stance in relation to it'. The long-term quantitative growth of media coverage might well be taken as the best indicator of how climate change has been consolidated as an issue of public concern. Yet at any given moment the quality of coverage matters very much, and it seems rash to discount the content of articles and focus only on their number. Jaspal and Nerlich (2012: 136) characterise coverage in 1988 as divided over the issue of how far climate change is caused by human activity, with majority, 'hegemonic' representations emphasising anthropocentric causes, and minority, 'polemical' constructions depicting it is a natural phenomenon. Over time, however, there have been variations not only in the quantity of coverage but also in its quality, in terms of the balance between different views of the causes of climate change, of the appropriate response to climate change, and indeed even of the reality of climate change. Researchers have found a sharp divergence between American and British news media in this respect, with the US press not only giving relatively less attention to the issue, but also allowing more room to climate sceptics (Boykoff and Rajan 2007: 208–9), and focusing less on solutions and responses to climate change as compared to the UK (Nerlich *et al.* 2012).

For Boykoff and Boykoff (2007: 1191), many of the quantitative and qualitative problems with climate change coverage stem from routine journalistic norms: they argue that 'professional, well-intentioned behavior can actually decrease the possibility of precise, proper, and pressing climate-change coverage'. In terms of quantity, what they call 'first-order' norms of personalisation, dramatisation and novelty determine what counts as a newsworthy story, so that if events are judged not to meet these criteria 'the chances for extensive, in-depth coverage of this environmental problem are diminished' (Boykoff and Boykoff 2007: 1200). This line of argument about selection criteria or 'news values' has a long pedigree, stretching back to Johan Galtung and Mari Ruge's seminal 1965 essay on the topic. There is also a precedent for Boykoff and Boykoff's 'second-order' norms, of 'authority-order' and balance. The former refers to the fact that 'journalists tend to primarily, and sometimes solely, consult authority figures – government officials, business leaders, and others – who reassure the public that order, safety, and security will soon be restored'; while the latter means that when there are competing authoritative claims about a news event journalists strive to give equal attention to different sides (2007: 1193).

Examining US news coverage from 1988 to 2004, Boykoff and Boykoff highlight a number of occasions when there was 'overt contestation' between different sources accessed by journalists, resulting in coverage that presented the views of 'duelling authorities' in a balanced way (2007: 1193). As noted above, this is how they account for increases in coverage in 1990, 1992, 2001–2 and 2004: occasions when the quantity of coverage rose but not, they argue, the quality. Relying on 'authoritative sources' who are 'duelling', and then trying to report their different perspectives in a balanced way, has meant that 'balance' has effectively led to 'bias':

> By operating in accordance with widely accepted journalistic norms, influential mass-media newspaper and television sources in the United States have misrepresented the top climate scientific perspective, and thus have perpetrated an informational bias regarding anthropogenic climate change.
>
> (Boykoff and Boykoff 2007: 1191)

Balanced reporting, they argue, misleadingly suggests an equivalence between 'thousands of the world's most reputable climate-change scientists who vigorously engage the process of peer review' and 'a few dozen naysayers who generally have not had their skeptical assertions published in peer-reviewed publications' (2007: 1193). It results in 'informational bias' because it creates 'an aura of scientific uncertainty' where none really exists, and this misleading impression of uncertainty is also 'a powerful political tool' which has 'helped to create space for the US government to defray responsibility and delay action regarding climate change' (2007: 1193, 1201).

This argument about the limitations of journalistic 'balance' recalls earlier critiques of the news media, particularly those made by Daniel Hallin in the 1980s and Stuart Hall in the 1970s. Both of these authors describe how in practice balance operates only within certain limits, delineating spheres of 'consensus', 'legitimate controversy' and 'deviance' (Hallin 1989: 117); or in Hall's (1973: 88) terms, of 'consensus', 'toleration' and 'dis-sensus or conflict'. Both writers are also sensitive to the way that what counts in the media as 'legitimate' or 'reasonable' disagreement over an issue can change over time, depending on the degree of elite consensus.[4] This critical understanding of the limits of 'balance' again entails 'indexing': it is journalists' reliance on influential official sources (in Hall's terminology, 'primary definers') which means that the scope of debate in the media tends to track the contours of elite concerns. It bears repetition that this has also been the case with regard to climate change, as is evident, indeed, from Boykoff and Boykoff's (2007: 1193) point about journalists' reliance on elite sources.

Their 2007 article extended a previous study (Boykoff and Boykoff 2004) which looked at US coverage up to 2002, but which also gave more detail about the quantities of balanced coverage. Between 1988 and 2002, a little over half of coverage (52.65%) was balanced regarding 'anthropogenic contributions to global warming'. Around a third of coverage (35.29%) was of a type which 'most closely mirrored the scientific discourse' by presenting both sides of the debate but emphasising the anthropogenic causes of climate change. A roughly equal amount of stories presented only one view – 6.18% questioning anthropogenic contributions, and 5.88% exclusively reporting anthropogenic contributions to global warming (Boykoff and Boykoff 2004: 129). They also examined coverage regarding action in response to climate change. Here, a much greater proportion – 78.2% – was balanced in terms of attention to 'courses of action that ranged from cautious to urgent and from voluntary to mandatory', and again roughly equal attention was given to each pole of the debate – 11.17% exclusively focusing on 'voluntary and cautious' policy responses, and 10.63% favouring 'immediate and mandatory' action (Boykoff and Boykoff 2004: 131). These proportions were not constant over the period, however: they note that 'in 1988 and 1989, the vast majority of coverage emphasized anthropogenic contributions to global warming' and 'concentrated on immediate and mandatory action' (2004: 130–1). Boykoff and Boykoff see this in terms of coverage 'mirroring the scientific discourse of the time', but it is not as if journalists consulted only scientists when sourcing news stories in the late 1980s. Rather, it would be more accurate to say that coverage mirrored greater elite consensus on the issue in the late 1980s and that it was only when political debate over climate change emerged strongly in the early 1990s that 'balance became a common feature of the journalistic terrain' (2004: 131).

Even more significant is what happened next. In a follow-up study, Boykoff (2007) examined coverage of climate change in the US and UK prestige press from 2003 to 2006. As noted above, coverage quadrupled in the UK and increased by two-and-a-half times in the US during this period, in response to high-profile political initiatives such as the 2005 G8 meeting, the publication of the *Stern Review* in 2006 and the release of *An Inconvenient Truth* the same year (Boykoff 2007: 472–3).

This was, in other words, a high point of elite consensus on climate change, and coverage followed suit, not only quantitatively, but also in terms of its quality. In the US press, balanced coverage of anthropogenic climate change fell from 36.59% in 2003 to just 3.3% in 2006. Articles depicting anthropogenic contributions to climate change as 'negligible' accounted for only 2.44% of coverage in 2003 and disappeared altogether in 2004–06 (Boykoff 2007: 474). In the UK, the already tiny amount of balanced coverage in 2003 (1.67%) fell to 0.41% by 2006. By the end of the period, 96.7% of US coverage and 99.59% of UK coverage emphasised anthropogenic contributions to climate change.[5] For Boykoff (2007: 471), such coverage in 'prestige' newspapers is important because they are 'major influences on policy discourse and decision making at national and international levels', and the problem with balanced coverage was that it created 'spaces for policy actors to defray responsibility and delay action'. Yet if, as we have argued, coverage of climate change has followed elite concerns, and has been indexed to the shifting lines of political debate on the issue, then this puts the very high degree of consensus achieved by 2006 in a rather different light.

In the past, when critics argued that the role of the journalist in relation to 'consensus values' was to 'serve as an advocate and celebrant' (Hallin 1989: 117), the point was not to applaud and promote such a role. Equally, drawing attention to the sphere of 'legitimate controversy' within which journalistic balance routinely operates was not intended as an argument for limiting it still further and excluding a greater number of views as 'deviant'. Rather, the concern of scholars in the past was to critique taken-for-granted consensus assumptions, to challenge the narrow boundaries placed on debate in the mainstream media, and to question the way that some views were marginalised or simply ignored. In relation to climate change coverage, however, researchers' assumptions tend to be very different. In seeking to critique the barriers to 'precise, proper, and pressing climate-change coverage' (Boykoff and Boykoff 2007: 1191), the idea seems to be that 'proper' coverage would be unbalanced and one-sided: it would exclude any 'contrarian' views of the causes or extent of climate change, and would only discuss 'immediate and mandatory' policy responses. Debate and disagreement are largely understood to be illegitimate because only one side of the debate is seen as a political perspective, while the other side is simply 'scientific'. As Boykoff and Boykoff (2007: 1202) put it, journalistic norms and the news choices that flow from them play 'a crucial role in the failure of the central messages in the generally agreed-upon scientific discourse to transmit successfully into US-backed international policy to combat global warming'. Where those messages succeed in coming through in close to 100% of coverage, by this logic one would have to conclude that journalistic norms have been suspended in order to allow a direct transmission of scientific discourse.

Conclusion: misunderstanding the media

There are a number of problems with the perspective outlined above. First, it proposes an incoherent understanding of the news media. It is logically inconsistent to argue on the one hand that coverage of climate scepticism demonstrates elite influence on the media, but to maintain on the other that if there is evidence of media consensus in favour of environmentalist views then that is simply the proper transmission of scientific knowledge. Researchers seem unwilling to acknowledge the extent to which political leaders and other elite figures have positively promoted the issue of climate change, and to recognise that this has then been echoed in media coverage. It is notable, for example, that British MPs on the Parliamentary Science and Technology Committee have voiced similar criticisms to Boykoff, censuring the BBC for its 'false balance' in coverage of climate change (Vidal 2014). As we have seen, work in this area tends to skirt around the question of indexing, even as it accumulates evidence of its consistent operation. To interpret the virtual disappearance of balanced coverage described by Boykoff as evidence of journalists accurately conveying scientific knowledge ignores not just the indexing hypothesis, but the entire tradition of critical media scholarship and its investigation of the relationship between mainstream media and centres of political and social power. Much of the research on media and climate change only acknowledges the critical tradition as an explanation for why climate 'sceptics' and 'deniers' feature in media coverage, while the rise and consolidation of environmental issues in the mainstream media is treated as an eternal puzzle.

Second, the argument that balanced media coverage of climate-sceptic views (or indeed, of cautious policy responses) endows them with undue legitimacy suggests both an unusually censorious attitude and a naïve view of the relationship between science and politics. Researchers are effectively seeking to police the boundaries of what should count as acceptable opinion, on the grounds that environmentalist perspectives are based on science and should therefore be protected against the political and self-interested claims of climate-scepticism. This is to ignore the way that scientific and political discourses around climate change have been closely intertwined in terms of both international regulation (Franz 1997, Miller 2004) and national policy (Mahoney and Hulme 2016), ever since it became a political issue in the 1980s. Even if one were to ignore this, the notion that scientific findings must be protected against views that challenge them implies a decidedly peculiar understanding of scientific method and a highly restrictive stance toward public debate. In this respect, media research has contributed to a wider effort not only to de-legitimise but even to criminalise disagreeable or controversial perspectives on climate change. At the height of the consensus described by Boykoff, some activists and campaigners started to call for the prosecution of climate denial as a criminal offence.[6]

The logical next step – of arguing that media reporting of climate denial is also criminal – has been taken by, for example, *Truthout*'s Thom Hartmann, who also criticises mainstream journalism's 'false balance' in reporting the views of 'wackos who are often the hired hands of the fossil fuel industry' alongside those of 'the top climate scientists on earth' (Hartmann 2014). Appealing to the courts looks attractive to some campaigners and advocates precisely because it seems to have the potential to put the issue beyond political debate. British law professor Philippe

Sands QC, for instance, has argued that a ruling by the International Court of Justice would 'settle the scientific dispute' about climate change. Noting that a 'broad emerging consensus' on the anthropogenic causes of climate change and on the policies necessary to address it is still 'subject to challenge in some quarters, including by scientifically qualified, knowledgeable and influential individuals', Sands argues that 'the courts could play a role here in finally scotching those claims' (Sands 2015). It would be a strange kind of science that could be 'settled' by a judicial ruling, and it seems peculiarly unscientific to propose that the authority of judges should silence the claims of people who are 'scientifically qualified' and 'knowledgeable', but in any case it is clearly a flawed proposal because it simply reproduces in another form the problem it seeks to solve. Invoking 'The Science' is already supposed to be a way to quash objections and produce consensus, and it seems unlikely that invoking 'The Law' instead would prove any more successful.

When elite opinion has been divided on the issue of climate change, this has indeed resulted in the sort of balanced coverage featuring 'duelling' sources which Boykoff and Boykoff (2007: 1193) describe. Yet equally, more one-sided coverage such as in the late 1980s or mid 2000s cannot be understood as simply a straightforward transmission of scientific discourse into public debate, but rather reflects the fact that these were high-points of elite – and therefore media – consensus on the issue. The assumption that climate change has had to be forced on to media and elite agendas from below has made most researchers strangely blind to the operation of indexing in this area of coverage. Sophisticated model-building and a search for obscure 'behind-the-scenes' factors influencing fluctuations in news coverage should be abandoned in light of the fact that over time climate change has become a prominent and seemingly permanent feature of mainstream media and political discussion. If researchers wish to understand how the issue is inflected in news coverage at particular moments, the best starting point is to look at the broader (post-) political context in which the media operate.

The two problems affecting research on media coverage identified above are symptomatic of a third underlying issue: this approach adopts normative goals – to facilitate the transmission of scientific consensus into policy via the media, and to prevent the circulation of sceptical views – which lead its advocates into what Pieter Maeseele (2015: 391) calls 'the post-political trap'. As he observes, much of the research on media coverage of climate change 'primarily evaluates public discourse on the extent to which it either contributes to communicating a scientific consensus or to achieving a social consensus', and in doing so researchers 'not only actively contribute to the *de-politicization* of climate change, but actually presume that (an "unjustified") *politicization* is the problem to overcome' (Maeseele 2015: 393–4, original emphasis). As we saw in the Introduction, a number of critics have argued that simply promoting consensus, casting climate change as an issue that is 'above' or 'beyond' political contestation, tends to result in disengagement and a lack of social or political change. In the work of these critics, there is finally an acknowledgement that climate change is a thoroughly mainstream issue, taken up willingly and enthusiastically by elite figures and powerful institutions. The problem, for critics of the post-political, is not that climate change is ignored or marginalised but rather that it is incorporated into the mainstream via 'hegemonic technomanagerial discourses' which serve only to heighten a 'sense of political powerlessness' (Carvalho and Peterson 2012: 7, 2). This problem is addressed in the next chapter.

Notes

1 The 'political economy' approach is perhaps most closely associated with the work of Edward Herman and Noam Chomsky (2002), though that in turn is part of a longer tradition which derives the key characteristics of news media from the fact that most of them are capitalist enterprises (for example, Murdock and Golding 1974, Schiller 1976). The distinction between 'culturalist' and 'structuralist' approaches comes from the early days of British Cultural Studies when a division was drawn between a humanist and historical tradition, exemplified in the work of Raymond Williams and E.P. Thompson, that emphasised agency and 'lived experience'; and a theoretically driven anti-humanist paradigm, of which Louis Althusser was the most influential exponent, that emphasised the 'determinate conditions' and discursive regimes through which 'experience' was structured (see, for example, Hall 1980).

2 *Time*'s 2 February 1970 edition is available at: http://content.time.com/time/maga-zine/0,9263,7601700202,00.html. For a discussion of news magazine coverage of the environment in this period, including some of the 'behind-the-scenes' factors promoting it, see Gibson 2002: 71–3.

3 Archived at: http://webarchive.nationalarchives.gov.uk/+/http:/www.hm-treasury.gov. uk/independent_reviews/stern_review_economics_climate_change/stern_review_ report.cfm.

4 In the case that Hallin (1989) examines – coverage of Vietnam – opposition to the war was initially treated as 'deviant' and was marginalised as unreasonable and extreme, while critical discussion was narrowed to the bounds of 'legitimate' controversy (such as over the tactical execution of the war, as opposed to more fundamental questions about its legitimacy). As the elite consensus in favour of the war broke down, the boundaries of debate shifted, allowing some of what was formerly 'deviant' opinion to become part of 'legitimate' debate.

5 Painter and Gavin (2015: 7) dispute Boykoff's methodology and findings, conducting their own study and finding somewhat higher levels of climate-scepticism in UK press coverage. In the newspapers they examined, 7% of articles in 2007 'included skeptical voices', and in the two papers with the highest number of such articles that rose to 13%.

6 Comparing 'climate denial' with 'Holocaust denial', British campaigner Mark Lynas suggested in 2006 that there should be 'international criminal tribunals [for] those who will be partially but directly responsible for millions of deaths from starvation, famine and disease' (Lynas 2006). Drawing a similar comparison, David Roberts, a writer for the climate website *Grist*, suggested 'some sort of climate Nuremberg' for deniers (Roberts 2006). Canadian environmentalist David Suzuki has repeatedly called for political leaders to be jailed for climate change denial (Offman 2008, Heck 2016). James Hansen has told a Congressional committee that companies funding 'contrarian' views are guilty of 'crimes against humanity' (Nastu 2008), and US law professor William Tucker has made a legal argument for prosecution (Tucker 2012, 2012–13).

References

Anderson, Alison (2009) Media, politics and climate change: Towards a new research agenda, *Sociology Compass*, 3 (2): 166–82.

Anderson, Alison (2014) *Media, Environment and the Network Society*. Basingstoke: Palgrave.

Bennett, W. Lance (1990) Toward a theory of press-state relations in the United States, *Journal of Communication*, 40 (2): 103–25.

Boykoff, Maxwell T. (2007) Flogging a dead norm? Newspaper coverage of anthropogenic climate change in the United States and United Kingdom from 2003 to 2006, *Area*, 39 (4): 470–81.

Boykoff, Maxwell T. (2009) We speak for the trees: Media reporting on the environment, *Annual Review of Environment and Resources*, 34: 431–57.

Boykoff, Maxwell T. and Jules M. Boykoff (2004) Balance as bias: Global warming and the US prestige press, *Global Environmental Change*, 14: 125–136.

Boykoff, Maxwell T. and Jules M. Boykoff (2007) Climate change and journalistic norms: A case-study of US mass-media coverage, *Geoforum*, 38: 1190–1204.

Boykoff, Maxwell T. and S. Ravi Rajan (2007) Signals and noise: Mass-media coverage of climate change in the USA and the UK, *EMBO Reports*, 8 (3): 207–11.

Boykoff, Maxwell T. and J. Timmons Roberts (2007) *Media Coverage of Climate Change: Current Trends, Strengths, Weaknesses*, United Nations Development Programme *Human Development Report* background paper, http://hdr.undp.org/sites/default/files/boykoff_maxwell_and_roberts_j._timmons.pdf.

Carvalho, Anabela and Jacquelin Burgess (2005) Cultural circuits of climate change in UK broadsheet newspapers, 1985–2003, *Risk Analysis*, 25 (6): 1457–69.

Carvalho, Anabela and Tarla Rai Peterson (2012) Reinventing the Political, in Anabela Carvalho and Tarla Rai Peterson (eds.) *Climate Change Politics: Communication and Public Engagement*. Amherst, NY: Cambria Press.

Downs, Anthony (1972) Up and down with ecology – the 'issue-attention cycle', *The Public Interest*, 28: 38–50.

Franz, Wendy E. (1997) *The Development of an International Agenda for Climate Change: Connecting Science to Policy*. Environment and Natural Resources Program Discussion Paper E-97-07, Kennedy School of Government, Harvard University, August.

Galtung, Johan and Mari Holmboe Ruge (1965) The structure of foreign news, *Journal of Peace Research*, 2 (1): 64–90.

Gibson, Donald (2002) *Environmentalism: Ideology and Power*. Huntington, NY: Nova Science Publishers.

Hall, Stuart (1973) A World at One with Itself, in Stanley Cohen and Jock Young (eds.) *The Manufacture of News*. London: Constable.

Hall, Stuart (1980) Cultural studies: Two paradigms, *Media, Culture & Society*, 2 (1): 57–72.

Hall, Stuart, Chas Critcher, Tony Jefferson, John Clarke and Brian Roberts (1978) *Policing the Crisis*. London: Macmillan.

Hallin, Daniel C. (1989) *The 'Uncensored War': The Media and Vietnam*. Oxford: Oxford University Press.

Hammond, Philip (2007) *Media, War and Postmodernity*. London: Routledge.

Hansen, Anders (2015) News Coverage of the Environment: A Longitudinal Perspective, in Anders Hansen and Robert Cox (eds.) *The Routledge Handbook of Environment and Communication*. Abingdon: Routledge.

Hartmann, Thom (2014) The mainstream media's criminal climate coverage, *Truthout*, 26 February, www.truth-out.org/opinion/item/22123-the-mainstream-medias-criminal-climate-coverage.

Heartfield, James (2008) *Green Capitalism*. London: Open Mute.

Heck, Alexandra (2016) David Suzuki thinks Stephen Harper should serve prison time for 'wilful blindness' to climate change, *National Post*, 2 February, http://news.nationalpost.com/news/canada/david-suzuki-thinks-stephen-harper-should-serve-prison-time-for-inaction-on-wilful-blindness-to-climate-change.

Herman, Edward S. and Noam Chomsky (2002) *Manufacturing Consent* (Second Edition). New York, NY: Pantheon.

Hilgartner, Stephen and Charles Bosk (1988) The rise and fall of social problems: A public arenas model, *American Journal of Sociology*, 9 (1): 53–78.

Jaspal, Rusi and Brigitte Nerlich (2012) When climate science became climate politics: British media representations of climate change in 1988, *Public Understanding of Science*, 23 (2): 122–41.

Lester, Libby (2010) *Media and Environment*. Cambridge: Polity.

Lynas, Mark (2006) Climate denial ads to air on US national television, *MarkLynas.org*, 19 May, https://web.archive.org/web/20080512154243/http://www.marklynas.org/2006/5/19/climate-denial-ads-to-air-on-us-national-television.

Maeseele, Pieter (2015) Beyond the Post-Political Zeitgeist, in Anders Hansen and Robert Cox (eds.) *The Routledge Handbook of Environment and Communication*. Abingdon: Routledge.

Mahoney, Martin and Mike Hulme (2016) Modelling and the nation: Institutionalising climate prediction in the UK, 1988–1992, *Minerva*, 54 (4): 445–70.

Mazur, Allan (1998) Global environmental change in the news: 1987–90 vs 1992–6, *International Sociology*, 13 (4): 457–72.

McCombs, Maxwell E. and Donald L. Shaw (1972) The agenda-setting function of mass media, *Public Opinion Quarterly*, 36 (2): 176–87.

Mermin, Jonathan (1999) *Debating War and Peace*. Princeton, NJ: Princeton University Press.

Miller, Clarke A. (2004) Climate Science and the Making of a Global Political Order, in Sheila Jasanoff (ed.) *States of Knowledge: The Co-Production of Science and Social Order*. London: Routledge.

Murdock, Graham and Peter Golding (1974) For a Political Economy of Mass Communications, in Ralph Miliband and John Saville (eds.) *The Socialist Register 1973*. London: Merlin.

Nastu, Paul (2008) James Hansen: Try fossil fuel CEOs for 'high crimes against humanity', *Environmental Leader*, 24 June, www.environmentalleader.com/2008/06/24/james-hansen-try-fossil-fuel-ceos-for-high-crimes-against-humanity/.

Nerlich, Brigitte, Richard Forsyth and David Clarke (2012) Climate in the news: How differences in media discourse between the US and UK reflect national priorities, *Environmental Communication*, 6 (1): 44–63.

Offman, Craig (2008) Jail politicians who ignore climate science: Suzuki, *National Post*, 6 February, www.nationalpost.com/news/story.html?id=290513.

Painter, James and Neil Gavin (2015) Climate skepticism in British newspapers, 2007–2011, *Environmental Communication*, 10 (4): 432–52.

Pepermans, Yves and Pieter Maeseele (2014) Democratic debate and mediated discourses on climate change: From consensus to de/politicization, *Environmental Communication*, 8 (2): 216–32.

Roberts, David (2006) The denial industry: An excerpt from a new book by George Monbiot, *Grist*, 20 September, http://grist.org/article/the-denial-industry/.

Robinson, Piers (2002) *The CNN Effect*. London: Routledge.

Sands, Philippe (2015) Climate Change and the Rule of Law: Adjudicating the Future in International Law, United Kingdom Supreme Court, 17 September [transcript: www.kcl.ac.uk/law/newsevents/climate-courts/assets/CLIMATE-CHANGE-INT-COURTS-17-Sept.pdf; video: www.youtube.com/watch?v=eef1tK8mtEI].

Schiller, Herbert I. (1976) *Communication and Cultural Domination*. White Plains, NY: International Arts and Sciences Press.

Tucker, William C. (2012) Deceitful tongues: Is climate change denial a crime?, *Ecology Law Quarterly*, 39 (3): 831–92, http://scholarship.law.berkeley.edu/cgi/viewcontent.cgi?article=2014&context=elq.

Tucker, William C. (2012–13) The big lie: Is climate change denial a crime against humanity?, *Interdisciplinary Journal of Human Rights Law*, 7 (1): 90–125.

Ungar, Sheldon (1992) The rise and (relative) decline of global warming as a social problem, *Sociological Quarterly*, 33 (4): 483–501.

Vidal, John (2014) MPs criticise BBC for 'false balance' in climate change coverage, *Guardian*, 2 April, www.theguardian.com/environment/2014/apr/02/mps-criticise-bbc-false-balance-climate-change-coverage.

Williams, Raymond (1977) *Marxism and Literature*. Oxford: Oxford University Press.

3
GREEN CONSUMPTION, LIFESTYLE JOURNALISM AND MEDIA ADVOCACY

Some critics have suggested that there needs to be a new agenda in media and climate change research – one concerned, not with the 'proper' communication of consensus as discussed in Chapter 2, but with the ways that discourses about climate change promote politicisation or depoliticisation. Anabela Carvalho and Tarla Rai Peterson (2012: 8), for example, observe that top-down climate change communication initiatives usually involve little more than 'persuading individual citizens to voluntarily modify some aspect of their energy-related behavior or accept some policy proposal that typically stays within the limits of existing economic and institutional structures'. In this way, members of the public are 'addressed as individual consumers', so that 'responsibility for climate change is individualized and the political realm is reduced to lifestyle choices' (2012: 8). The line of enquiry opened up by Carvalho and Peterson is further elaborated by Yves Pepermans and Pieter Maeseele, who call for a reorientation of research on media and climate change, away from the narrow evaluation of media performance in terms of the 'proper communication of (the scientific consensus on) climate science', and toward a concern with the 'discursive strategies underlying processes of politicization and depoliticization' in media and public discourse (Pepermans and Maeseele 2014: 219, 223).

This chapter takes up this new agenda, focussing on how 'green lifestyle journalism' promotes the sort of depoliticised responses that Carvalho and Peterson highlight, comparing examples from the BBC and the *Guardian*. It examines how these two news organisations have attempted to engage their audiences as potentially green consumers by offering models of 'ethical', environmentally-friendly behaviour, but in different ways. For the BBC, which has a long tradition of journalistic balance and impartiality, the depoliticisation of climate change could be understood as a sort of institutional requirement. For the *Guardian*, in contrast, with its strong reputation as a committed and campaigning newspaper, climate change is an issue that has been understood to demand an explicitly partisan, advocacy stance. The media, it is argued below, are not 'to blame' in any straightforward sense for depoliticisation, but they do offer representations that dramatise and enact therapeutic conceptions of 'agency', even when they are attempting to be overtly political.

Climate change as ideology

A good starting point is Peter Berglez and Ulrika Olausson's (2014: 57) suggestion that climate change can be understood as an 'ideological discourse' which works to 'preclude…critical questioning of the predominant socio-political order'. Although Berglez and Olausson's research is about everyday public attitudes rather than news coverage, their investigation of the 'ideological processes that generate the consensual, post-political condition of climate change' (2014: 60) offers a useful framework for analysing climate change as ideology in media discourse.

According to Berglez and Olausson (2014: 56), there are three features of the 'consensual discourse' which constitute climate change as a post-political issue: '1. belief in a "climate threat," 2. personal experiences of a "climate threat," and 3. integration of a "climate threat" into everyday-life practice'. This condensed formulation perhaps needs a little unpacking. Their first point is that 'naturalized beliefs in climate science' operate as 'an abstract authoritative voice', as 'something beyond rational questioning that does not seem to need further evidence' (2014: 69, 60–1). Though ostensibly based on science, climate change as ideology operates in an unscientific register, much in the manner discussed at the end of Chapter 2, whereby 'elements of scientific uncertainty have been washed away, and the scientific hypotheses become univocal and consensual truths' (2014: 61).

Second, they argue that people further naturalise these abstract truths by 'narrativizing' them: incorporating them into their own life-narratives by relating them to personal experience, for example through interpreting weather events as evidence of climate change (2014: 63–4). Berglez and Olausson argue that this is typically done in a way that does not involve 'strong' or 'passionate' emotions – which, they think, could transform such personal narratives into the basis for political engagement. Instead, there is at most a sense of 'sadness and nostalgia about the loss of features of the flora and fauna which have "always been there"' (2014: 65).

Finally, their third point is that these naturalised beliefs become incorporated into 'the practical routines of everyday life, including language', so that:

> one gradually gets drawn into a discourse that already seems to have become 'standard', a socially accepted way of talking and thinking about climate change, and an established trend somewhere else: in the media, in commercials, at work, in education, in everyday parlance, and in politically correct conversation. This discourse is endowed with certain standard environmental concepts such as recycling and organic food, but also with related practices such as the 'right way' of cleaning, consuming, driving, and so forth.
>
> (Berglez and Olausson 2014: 65)

The more that climate change is understood in this way, as requiring a kind of etiquette of correct behaviour, 'the more protected it becomes from critical questioning' (2014: 65). Instead, they argue, 'anxiety and fear become normal aspects of everyday subjectivity, generating constant micro-actions, such as recycling' (2014: 68). This form of what Berglez and Olausson (following Isin 2004) characterise as 'neurotic' subjectivity does not lead to any 'radical political transformation of present society' or 'pose any challenge to the capitalist system', but instead has the opposite effect of 'reinforcing this very order' (2014: 68).

Matthew Paterson and Johannes Stripple (2010: 342) observe that there is a 'long heritage' of framing environmental issues in terms of 'individual practice and virtue' and a correspondingly long tradition of critiquing this individuation, particularly as it pertains to 'the neoliberal era, when dominant ideologies attempted to displace responsibility for all sorts of problems from the state and onto individuals'. One could add that there is another contribution to this debate with an equally substantial history: writers acknowledging the critique of neoliberalism's 'responsibilisation' of the individual but then minimising or discounting it. Paterson and Stripple themselves, indeed, conclude that it is not possible to respond effectively to climate change without 'an intensive, managerial (and self-managerial) effort', which would shape not only the behaviour of subjects 'but their internal rationalities, identities, what they fundamentally regard as "normal" behaviour' (2010: 359).

Paterson and Stripple's let's-get-with-the-programme approach to neoliberal governance is unusual in the literature, but their insistence that it does not entail any sort of 'depoliticisation strategy' is quite typical (2010: 359). Writing on 'green lifestyle' television programming, for example, Tania Lewis notes that:

> the current preoccupation within contemporary culture with the lifestyle choices and consumption practices of ordinary people can be linked to the broader emergence of a hegemonic culture of devolved self-governance… [displacing] questions of social responsibility away from government and corporations onto individuals and their lifestyle 'choices', reflecting a growing 'ethicalization of existence'.
>
> (Lewis 2008: 227)

In the next breath, however, she argues that this should not be understood in terms of a decline or death of politics, but more as the opening up of a new and different sphere of political activity:

> Rather than assuming we have witnessed the demise of a civic-minded public, a number of critical scholars have been concerned with thinking about the ways in which civic concerns and debates are increasingly enacted in different ways and in different sites from those offered up in conventional accounts.
>
> (Lewis 2008: 228)

In this scenario, depoliticisation is simply not a problem. While people may be 'increasingly cynical about the benefits of engaging with formal political processes such as voting', this is just because politics has moved elsewhere, into a 'more micro-political, lifestyle-based mode of civic agency and citizenship', where everyday consumption becomes 'a site through which ordinary people invest in ethical, social and civic concerns' (2008: 228). In this perspective, there can be negative and positive versions of ethical responsibilisation: whereas neoliberal consumer-citizenship promotes 'materialism, excess and selfish individualism', in the 'ethical' version there is 'a broader focus on questions of impact, both in terms of social relations and the environment', so that the 'consumer-citizen is…positioned as an active player in social relations' (Lewis 2008: 238).

Contributors to this debate often take their cue from the work of Nikolas Rose (1999: 166), who argues that for today's 'responsibilised' subjects, 'Citizenship is no longer primarily realized in a relation with the state, or in a single "public sphere", but in a variety of private, corporate and quasi-public practices from working to shopping'. Despite registering the ways in which the ethical responsibilisation of the citizen/consumer can act as 'a recoding of strategies of social discipline and morality', extending government ever more deeply into the 'micro-management' of areas of life formerly understood as private, Rose nevertheless welcomes the 'infusion of ethical discourse into politics', because instead of a modernist 'will to truth' it allows 'the possibility of opening up the evaluation of forms of life and self-conduct to the difficult and interminable business of debate and contestation' (1999: 192–3). Developing this argument in relation to ethical consumerism, Clive Barnett and his co-authors argue that it encourages people to 'recognize themselves as bearing certain types of *global* obligation by virtue of their privileged position as consumers, obligations which in turn they are encouraged to discharge in part by acting as consumers in "responsible" ways' (Barnett *et al.* 2011: 13, original emphasis). In this way, the 'consumer choice' involved in ethical consumption is less about 'unfettered hedonistic self-interest', and more about 'the responsibilities of self-monitoring and caring for others' (2011: 70).

In one sense it is true that politics has moved somewhere else, but in the process 'politics' has been re-defined and reinterpreted. Political action is now increasingly understood, not in a modernist sense of subjective agency working to transform the object world, but more as a project of work on the self. As one study of fair trade ethical consumerism puts it: 'While doing good is without doubt a central objective in practices of fairtrade consumption, the main concern in accounting for such practices is being good in terms of working towards an ideal self' (Varul and Wilson-Kovacs 2008: 8). This does not mean 'ideal' in a strong normative sense, although there is a normative dimension in terms of how being a 'good person' is understood, and, as we have noted, in terms of an etiquette of 'correct' behaviour. Rather, as Engin Isin's (2004: 232) notion of the 'neurotic citizen' who is (self-)governed through affect and emotion suggests, it is the process of 'working towards' which is key: the engagement with a process of reflexive work on the self.

Following Isin's (2004: 229) suggestion that 'As subjects settled into various ways of being environmental citizens by reusing, recycling, and reducing their consumption, the environmental discourse increasingly constituted them as neurotic citizens charged with saving the environment by these everyday habits', we can understand this as a therapeutic mode of governance. Writers looking for something positive in ethical consumerism are pleased to discover that it does not entail an 'image of the consumer as a purely self-interested individual' (Lewis 2008: 229), and nor does it focus narrowly on consumption as 'a locus of sovereignty and agency' (Barnett *et al.* 2011: 19). Yet contemporary modes of governance do not presuppose a competent, capable and calculating rational subject: rather they are addressed to vulnerable, emotional, 'neurotic' subjects, who are 'anxious, under stress and increasingly insecure' (Isin 2004: 225).

In the following two case-study examples, we explore these ideas by looking at how media organisations have attempted to engage audiences with climate change through the notion of ethical consumption. In the first example, from the BBC, the sort of depoliticisation described by Berglez and Olausson is both necessary for institutional reasons, and also possible because of the broader political context. Our focus here is on understanding how the news media have turned 'neurotic citizen' behaviours into infotainment. In the second example, from the *Guardian*, we look at a news organisation which is explicitly trying to politicise the issue and campaign around it, rather than to remain impartial and depoliticise climate change. Here, our interest is in how efforts to engage audiences in a more overtly political way, and to radicalise ethical consumption via a more thorough-going anti-consumerism, nevertheless tend to reproduce understandings of ethical/political agency as therapeutic and inward-focussed.

'Ethical Man' at the BBC

'Ethical Man' was a running feature on BBC Two's *Newsnight* programme from February 2006 to April 2007. The conceit of the series was that reporter Justin Rowlatt and his family would spend a year 'living ethically' by attempting to reduce their environmental impact. Towards the end of the project, the *Newsnight* segments were also combined and re-edited for an edition of the BBC1 current affairs programme *Panorama* (5 March 2007). Ethical Man was later 'reborn' for a further series of items for *Newsnight* from the USA in 2009, and Rowlatt also did a number of other reports on environmental issues, including two editions of the BBC Radio Four programme *Analysis* in 2010. Most of the later reports were closer to conventional journalistic pieces, but the Ethical Man format, particularly in its initial 2006–07 version, was quite different and strikingly novel – all the more so because it was part of the BBC's flagship current affairs programme.

Light-hearted and entertaining in tone, with much use of music, humorous headlines on *Newsnight*'s accompanying webpages ('Ethical Man's water retention', 'Ethical Man's wind problems'),[1] and occasional use of speeded-up film for comic effect, the format also drew on the conventions of reality TV genres. As Rowlatt reflected after compiling a selection of material for the *Panorama* programme:

I've been told to 'piss off' by my wife on national TV, I've been filmed urinating on my compost heap, I've scrumped fruit from my neighbours' gardens, filmed my own children crying and even had to smell a hippy's poo.

With all the rows and arguments it sometimes felt like we were making an episode of Wife Swap rather than a Panorama on how to tackle climate change.[2]

The entertaining style was perhaps emphasised due to a certain self-consciousness about incorporating 'green lifestyle journalism' – a genre which, as Geoffrey Craig (2016: 124) notes, is often seen as lightweight 'soft news' – into serious current affairs programming. Although certainly a departure for BBC journalism, it was also typical of its era: Jen Schneider and Glen Miller (2011: 470) note its relationship to the 'eco-stunts' and lifestyle experiments documented in films such as *Garbage: The Revolution Starts at Home* (2009), *Recipes for Disaster* (2008), and *No Impact Man* (2009); to 'consumer consciousness days' like Buy Nothing Day; and to the many books giving advice on 'green' consumption and 'ethical living' (for example Barclay and Grosvenor 2007, Gow McDilda 2007, Hailes 2007, Hickman 2008).

Perhaps the most immediate stylistic reference point was the emerging sub genre of what Lyn Thomas (2008) calls 'eco-reality' programming. Thomas discusses two examples from 2005 – *No Waste Like Home* (BBC2) and *The Real Good Life* (ITV) – which exemplify different approaches to green lifestyle television: programmes of the first type, she observes, are 'strongly didactic', offering a 'narrative of self-improvement', often under the guidance of an expert, and drawing on 'discourses of low and restored self-esteem which emanate from the well established domains of self-help and popular psychology'; whereas in the second type, 'relationships under pressure is the real theme', with people being placed in unusual and challenging situations (Thomas 2008: 693–4).[3] Ethical Man combined elements of both. There was an overarching makeover narrative centred on Rowlatt's efforts to become more 'ethical' and reduce the household 'carbon footprint', which was measured at the beginning and end of the project to produce 'before' (above average) and 'after' (reduced by 20%) results. But the series also strongly featured the on-screen relationship between Rowlatt and his partner, Bee (eventually given her own billing as 'Ethical Wife'), showing us the strains that the experiment put on their family life.

Laurie Ouellette and James Hay (2008: 471) have noted how the popularity of the makeover genre in US and UK TV is 'tied to distinctly "neoliberal" reasoning about governance and social welfare', focusing as it does on 'the diffusion of personal responsibility and self-enterprise as ethics of "good" citizenship'. When it comes to the genre's green variants, however, critics are more inclined to see this in a positive light. Thomas (2008: 696–7), for example, counters imagined critics who 'might argue that these narratives are part of the focus on consumer desires', maintaining instead that such programmes depict 'a turn to a domestic and local version of citizenship' which has 'potential for political mobilization'. Similarly, Lewis (2008: 230) takes it as a positive sign that green lifestyle programming is not concerned with 'the promotion of consumerism' but with 'ethical concerns around consumption and in particular with shaping the values and practices of viewer-consumers as responsible citizens'.

Ethical Man was not at all concerned with promoting consumerism, and nor did it take a strongly didactic approach. Instead, Rowlatt's affable persona as Ethical Man offered something similar to what Marilyn DeLaure (2011: 452) finds in *No Impact Man*: 'a more approachable and critically engaging template for every-day living, inspired by the comic fool'. The entertaining and amusing format was intended to elicit audience engagement (whether for the sake of the ethical issues or *Newsnight*'s viewing figures): from the first episode, viewers were encouraged to visit the programme website to give suggestions and relate their own experiences. The beginning of Ethical Man coincided with the BBC's first forays into social media, so the initial online format of a 'Diary of an Ethical Man' article with selected readers' letters soon gave way to an 'Ethical Blog', on which viewers could comment.[4] Judging by the published responses, although some viewers did not like the light-hearted style, the reaction was generally positive.

Ethical Man was an ambiguous device, simultaneously contrived and authentic; both distancing and highly personal. On one hand, Ethical Man was very clearly a constructed persona, distinct and separate from Rowlatt himself; on the other hand, the formula relied on giving viewers a glimpse of Rowlatt's private life, much in the manner of a fly-on-the-wall documentary. The series balanced two logically contradictory demands: it implicitly advanced a proposition about what would constitute an 'ethical' lifestyle, while also seeking to maintain the traditional impartiality of BBC journalism. The tension between these two was managed via the interplay between the public persona and the private person but, perhaps surprisingly, both the ethical proposition and the institutional impartiality were ultimately underwritten by the 'private' life of Rowlatt and his family rather than the overtly public journalistic discourse.

The separation between the reporter and the persona was signalled visually by the green suit he always wore, even in situations where it was incongruous (cycling, on the beach, in a muddy field), and which he composted in a 'burial' of Ethical Man in the final 2007 episode. The distance was also stated explicitly and emphatically from the outset. The first episode (22 February 2006) began with Rowlatt saying:

> I've just started on *Newsnight*, but my editor's already got an idea for me: do I want to turn my life completely upside down as part of an experiment in ethical living? Well, I've only just started on the programme so, eh, yeah, of course I do.
> But before this experiment starts, I've booked a holiday.[5]

Rowlatt is reluctant, slightly put-upon: ethical living does not 'come naturally'; he likes driving and would rather not give up his car; he relishes the chance to snatch a quick holiday in the Canary Islands before the project officially starts. He underlined the point in the first of his many online commentaries on the project: 'I want to be clear about one thing right from the start: it wasn't my idea to become *Newsnight*'s "ethical man"'.[6] Similarly, when *Newsnight* revived the strand for its 2009 US version, Rowlatt reminded viewers that he and his family had taken part in what he described as a 'bizarre BBC experiment' only because his editor had 'ordered' them to do so.[7]

This distancing was important in allowing a version of journalistic impartiality to operate: Rowlatt the private person, with his own individual views and opinions, was separate from Rowlatt the public persona who acted in accordance with a set of rules. In more conventional journalism, the public persona is that of the professional reporter: this is kept distinct from the private beliefs of the individual who occupies that role, and the rules are those of BBC impartiality.[8] In the scenario that Ethical Man established, however, the public persona was required to act, not 'impartially', but in an 'ethical' and 'environmentally-friendly' way. For this to work in terms of impartiality, it was important that the BBC itself was not seen to be motivated by any overt commitment to environmentalism: the motives of *Newsnight* editor Peter Barron in commissioning the series were not discussed, and although Rowlatt's producer, Sara Afshar, was a frequent off-screen presence, the only motivation imputed to her was the successful execution of the experiment. Rowlatt referred to her devising difficult tasks for him to attempt, saying for example: 'Sara's got an even tougher challenge for me next: on Monday she's going to take away my car, indefinitely',[9] or 'my producer Sara has plans for me…she wants me to explore the impact of food on the environment by going vegan'.[10]

The main institutional representative of the BBC, however, was Rowlatt himself, and paradoxically, it was his personal views, rather than his professional identity, that underwrote his ability to be impartial. Had the BBC chosen a reporter whose private views were those of a committed green, the result would not only have been very different, it would have been a major departure from the established style of BBC journalism.[11] When the series was repackaged for *Panorama* in 2007, the rationale for the experiment needed to be restated. This time, the idea was pitched – under the title 'Go Green or Else' – as an investigation of possible future UK government measures to compel citizens to adopt environmentally-friendly behaviours. Presenter Jeremy Vine introduced it by saying:

> Meet the Rowlatts. First we took away their car, then their holiday flights, and then their Sunday lunch, so you could enjoy their pain. But one day we could all be forced to go green, so watch, and weep. Yes, the bin police are coming and road pricing could well be hot on their heels. Now if you're the kind of person who wakes up every morning wanting to save the planet, you will already be doing your bit. But the rest of us may just need some persuasion. Tonight we've got a preview of what's to come, with a family who've been forced to go green for the whole year.[12]

Having been introduced as showing us what it might be like if everyone was 'forced to go green', Rowlatt began by emphasising his personal reluctance to do so: 'I'm no tree-hugging eco-warrior, so I knew what to do when I was told to go green or else – I jetted off to the Canaries before they tried to turn me into an ethical man'. In a contemporaneous *Guardian* interview (which described him as an '*accidental* green hero') to promote the programme, Rowlatt again underlined the point: 'My initial reaction was that it all sounded really tedious and worthy, frankly. A bit like being force-fed muesli' (Silver 2007). It had to be said again and again: partly because it was what made Ethical Man an entertaining proposition – viewers could enjoy Rowlatt's 'pain' – but also because his professed private scepticism about the whole project was essential to sustaining a tone of impartiality.

Given that the BBC could not be seen as openly and directly advocating an environmentalist agenda, the implicit proposition about what constituted ethically-correct behaviour had to be supported in some other way. This was tackled by outsourcing the authority to judge whether Rowlatt was being ethical. The most important figure in this respect appeared to be Professor Tim Jackson, director of the research group on Lifestyles, Values and Environment at the University of Surrey, introduced as 'an expert on the impact consumers…have on the environment'.[13] However, although it was Jackson who carried out the carbon audit at the beginning and end of the project and who therefore judged its success, he did not make any sort of argument in favour of environmentally-friendly behaviour. *Guardian* columnist Leo Hickman, whose similar year-long experiment in 'living ethically' was the immediate inspiration for the project (Silver 2007), also featured in the first episode (introduced as an author of books on ethical living rather than as a fellow journalist), offering some advice on the personal difficulty of making lifestyle changes, but again not making any argument about why such changes might be necessary or desirable.

Instead, it was Rowlatt's sister, Cordelia, also featured in the first broadcast, who was the most passionate advocate of green 'ethical' living. She generates her own electricity with a solar panel, grows vegetables and keeps hens, has given up her car, does not fly on holidays, refuses to buy food that is airfreighted in, and is considering installing a compost toilet in the garden. As Rowlatt remarked, 'I've got a pretty good example of ethical living in my own family'. Challenged by Rowlatt over whether her individual efforts really made a difference, Cordelia argued that 'the main thing that can change is if people change their behaviour…so the best that I can do is to change what I do and hopefully that'll have an impact on other people'.[14] She elaborated the argument in an open letter to Ethical Man for the BBC website:

> I believe that the damage to the environment caused by people could be reduced very quickly. All it would take is for everyone to concentrate on changing their own behaviour.
>
> Imagine how quickly the world would change if everyone started trying their best to live a 'green' lifestyle tomorrow. The more people who make an effort, the closer we are to that happening. Moreover, this is probably the only way to change things. If we wait for someone else (the government) to act, we will be waiting a long time, possibly for ever.[15]

Fittingly, Cordelia's explanation is premised on scepticism about formal politics. Just as BBC impartiality was underpinned by Rowlatt's professions of private scepticism about the project, so too the implied ethical correctness of environmentalism was communicated not overtly by the BBC as an institution, but by Rowlatt's family. His sister rarely featured, though: the more subtle but significant and consistent contribution in this respect was made by his wife, Bee.

While the format required the construction of the contrived Ethical Man persona, it equally relied on creating the impression that we were getting a candid and intimate portrait of Rowlatt's private life. Viewers who followed Rowlatt and his family over the year witnessed his emotional highs and lows, sometimes recorded in confessional video-diary segments, including moments which would usually be considered highly personal and private, such as Bee going to hospital (on foot, since they had given up the car) to give birth to their baby. Although there were sometimes moments of family tension, the emotional tone was generally light-hearted and comic. Some of the comedy derived from editorial decisions that placed Rowlatt in amusing situations (in the bath wearing a plastic shower cap, urinating in his garden at night); but it also derived from the dynamics of his personal relationship with his wife. A comparison with Hickman's ethical living reports for the *Guardian* is instructive. These also involved occasional domestic disagreements involving his partner, Jane:

> To say Jane wasn't exactly sold on the idea would be an understatement; in fact, the uncouth word she used to express her feelings rang loud in my ears for the next 12 months.…Jane did, slowly, come round to the idea – sort of.
>
> (Hickman 2014)

Hickman's reports often showed Jane to be reluctant, questioning the need to adopt 'ethical' behaviours such as vegetarianism, needing to be persuaded and coaxed, and sometimes forcing him to compromise.[16] In Ethical Man's case, however, the dynamic played out in reverse: it was Bee who tended to occupy the moral high ground. As Rowlatt reflected: 'A lot of the time I end up looking like a bit of a plonker…the joke tends to be on me and my wife's always calling me to account' (Silver 2007).

Their on-screen relationship sometimes recalled the familiar sit-com trope of the wiser, long-suffering wife and the less mature, hapless husband. This was set up in the first episode, when Bee teases him about whether he will be a good Ethical Man, remarks that he does not do the recycling unless she is present (because 'you know I won't come after you'), and ridicules him for using the car for a short trip to buy beer.[17] It is always Bee who occupies the high ground in relation to their 'ethical' lifestyle changes. In another early episode (in which Bee mocks his clumsy efforts to make their home more energy-efficient and pokes fun at his child-like fascination for a portable electricity meter), when Rowlatt switches their energy supplier to a carbon-neutral company she points out that she had tried to make the same switch a couple of years earlier but he had objected to the expense.[18] Similarly, when they face the prospect of giving up the family car, it transpires that Bee has 'wanted to get rid of it for a while' and is 'looking forward' to being without it.[19] She also reacts angrily to his decision to illustrate a report on aviation and carbon-offsetting by flying to Jamaica for the weekend, urging viewers to send him 'lots of really abusive emails'. [20]

Importantly, however, unlike her sister-in-law, Bee was not a green evangelist. In a commentary for the *Newsnight* website at the beginning of the experiment she observed that 'it's not a project we took on voluntarily' and professed her interest in its 'social and community implications' as being 'more compelling than some of the "green" challenges'. In a later post she also said that 'I'm with all the people who have written in to Ethical Man to say that being ethical isn't just about being green'.[21] Nevertheless, Bee expressed her satisfaction that 'suddenly Justin can't argue with me any more when I want him to be worthy':

> Previously, the impetus of our environmental efforts as a household had been largely down to me being a nag. I've always insisted on Fair Trade products, free range meat, and a weekly organic box delivery, but I have felt a bit of a voice in the wilderness.
>
> Now I'm relieved of that burden. And I confess to moments of glee when old domestic arguments (about shares in an oil company, about charity donations) are suddenly re-ignited, but this time with the full weight of *Newsnight* behind me. Bonus![22]

Bee later took to the comments section of Rowlatt's blog to ask about his oil shares, and eventually gave 20% of the money to charity while switching the remainder to a 'deep green' investment fund.[23] Bee Rowlatt's willingness to accept the changes demanded by the project may be partly explained by the fact that she also worked for the BBC as a producer, though this was never alluded to and she appeared in a personal capacity. It was as Rowlatt's wife, not as a fellow BBC employee, that she repeatedly held him to account for his 'unethical habits', thereby implicitly endorsing the ethical correctness of the lifestyle changes they undertook. Her ethical/green inclinations were presented, as in her comments above, primarily in terms of 'domestic arguments' or 'being a nag', rather than being couched in more public or political terms.

In articulating environmentally ethical behaviour as a matter of personal relationships rather than politics, the narrative function of 'Ethical Wife' was to facilitate the sort of personal 'narrativisation' of climate change identified by Berglez and Olausson.[24] As we have seen, the necessity of adopting environmentally-friendly behaviours appeared as an externally imposed demand (whether as an editorial directive or an anticipated government policy), much in the way that Berglez and Olausson (2014: 60–1) suggest the 'consensual truths' of climate change discourse operate as an 'abstract authoritative voice' which is 'beyond rational questioning'. Ethical Man did not directly address the science of climate change until December 2009, nearly four years after the first episode, when Rowlatt hosted an experiment in his kitchen, carried out by space scientist Maggie Aderin-Pocock, involving heating plastic bottles containing different concentrations of carbon dioxide, in front of an audience made up of viewers of the BBC's motoring show *Top Gear*.[25] Significantly, this was a moment when the consensus-building approach was being put to the test. Rowlatt observed that:

> We're in the last week of negotiations at the UN Climate Change Summit in Copenhagen – but here in Britain the opinion polls show that about half of us don't actually believe that man is causing global warming. So for tonight's *Newsnight*…I will be trying to prove the basic science of global warming using a very simple scientific experiment.[26]

In the original Ethical Man series, there was evidently felt to be no need to 'prove' anything. At the time, this no doubt appeared to be a reasonable assumption: at the end of the series in 2007, Rowlatt noted that it had coincided with 'a fundamental change in the nature of the national and international debate about global warming', which he saw as having happened because 'the evidence of global warming has become so strong' – although he also described how 'Politicians [had] taken up the issue like never before' and had engaged in a 'greener than thou competition'.[27]

The series translated the 'consensual truths' from an otherwise abstract voice of authority into something personal and meaningful, via a narrativisation centring on the minor domestic dramas of Rowlatt's family life. The series understood 'tackling climate change' as a moral imperative problematising virtually all the everyday actions of individuals, as illustrated through the lifestyle changes made by the Rowlatts: giving up the family car, minimising energy use around the home, conserving water, going vegan for a month, and so on. It thereby legitimated an etiquette of 'environmentally correct' behaviour focused on individual micro-actions, not through overt argument (which, given the BBC's traditionally 'impartial' style would have been much more difficult to handle), but through presenting viewers with an entertaining and engaging display of therapeutic self-reflexivity.

Living our *Guardian* values

The ethical living experiment undertaken by the *Guardian*'s Leo Hickman required none of the distancing contrivances of the BBC's Ethical Man, since there was no institutional need to maintain an impartial stance. Like Rowlatt, Hickman used what he called 'the "innocent abroad" approach', but the rationale for doing so was to make a campaigning stance – attempting to 'get more readers excited by weighty, important subjects such as social and environmental responsibility' – more appealing by avoiding 'coming across as preachy and worthy'.[28] Also like Rowlatt, Hickman has recounted how the year of ethical living was a task handed to him by his editor, but in his case it was not simply a surprise external imposition: the idea grew out of a 'conversation with my editor at the *Guardian* about those pangs of consumer guilt that wash over us, but upon which we rarely act' (Hickman 2014).

Much more earnest in tone, Hickman's account revealed his personal investment in the project: 'In recent years the guilt has started to nag at me increasingly, but I can't seem to transform that caring instinct somewhere inside me into any kind of meaningful action' (Hickman 2006: Foreword, no page number). Hickman's private views and his public role as reporter were not kept separate: the experiment was not only an expression of his personal beliefs, but was also promoted by the *Guardian* as embodying the newspaper's pro-environmentalist outlook. The auditor's statement in the *Guardian*'s 2004 *Living Our Values* report highlighted Hickman's columns as showing what had been achieved at 'the level of personal and domestic responsibility', noting that: 'This innovative approach is a good example of the organisation engaging in the messy business of turning ideals and values into practical day-to-day reality'.[29]

The practice of publishing an annual report on how far the paper was 'living its values' began the previous year, initially as a 'social, ethical and environmental audit' and then from 2007 as a 'sustainability report'. Green issues were a very prominent feature of these values from the outset: in the first edition of the audit, in 2003, the *Guardian* described itself as having been 'the leading newspaper in environmental reporting for many years', but also as going beyond simply reporting to 'campaign strenuously for individuals, companies and governments to take action and act more responsibly'.[30] The 2003 report also cited the director of Friends of the Earth, Tony Juniper, praising the paper's 'contribution to environmental awareness':

> It is difficult to overestimate the impact of the *Guardian* and *Observer*. The *Guardian* is certainly considered as the voice of progressive and sound environmental thinking both in the UK and in Europe. The feedback we get from the US is that its influence is growing there too through the *Guardian*'s websites. It is the industry journal for environmental groups, as people can rely on its coverage of the issues that matter most.[31]

However, Juniper went on to observe that while it was imperative for the *Guardian* to 'continue playing its absolutely vital role in sustaining the environmental debate', it was not yet leading the way in terms of making its own company operations environmentally friendly and needed to 'put sustainability into practice', for example via its policies on 'paper purchasing and energy use'. Reporting on environmental issues is not enough, even campaigning actively to engage readers with environmental issues is not enough: the *Guardian* also wants to make its own practices as an organisation align with the values it espouses, and to issue yearly reports on its progress.[32] As this suggests, the newspaper itself has been engaged in a project similar to Hickman's domestic self-auditing.[33]

Jo Littler (2009: 62–3) usefully draws attention to the way that such Corporate Social Responsibility (CSR) audits can be understood as exemplifying the rise of what Eva Illouz (2007: 6) calls the 'therapeutic emotional style' of contemporary capitalism. Illouz suggests that the 'therapeutic idea of "communication"' came to define 'the spirit of the corporation' in the twentieth century, and highlights how compliance with this 'imperative of communication' entails a need to 'engage in a fairly complex work of introspection' (2007: 18–19). Illouz is discussing understandings of what it means to be a 'good manager' here, but the idea could be extended to encompass what it means to be a 'good business'. In the *Guardian*'s case, this endeavour is no doubt entirely sincere: its annual audits entail extraordinarily detailed self-examination, including even such matters as whether biodegradable cleaning fluids are used in the office toilets.[34] Yet it has not been straightforward to translate this therapeutic exercise into a clear political stance, either for the paper itself nor its individual contributors.

Hickman's pangs of guilt prompted a kind of existential angst, leading him to confront 'life's Big Questions: Who am I? What am I doing with my life? Is this it? Am I a good person?' (2006: Foreword). Seeking answers in the life-project of 'ethical living' translated the Big Questions into a series of more mundane dilemmas which centred on trying to 'define what we meant by "a more ethical lifestyle"':

> if you have to choose, is Fair Trade coffee better than organic? Are disposable nappies OK to use at night if washable ones leak? Should I always give loose change to someone holding a collection box in their hand? In terms of pollution, should I get the bus or the train to work? Should I let my child watch television adverts?
>
> (Hickman 2004)

Questioning the meaning of an 'ethical lifestyle' is a recurrent motif of this genre of journalism. As we saw earlier, Bee Rowlatt raised similar questions, preferring 'social and community' issues, and supporting viewers who wrote in to say that 'being ethical isn't just about being green'. In their online responses, viewers asked 'Why is Ethical Man solely concerned with the environment?', and 'whose ethics are we trying to live by?', for example, or suggested that 'it would be considerably more ethical to simply maximise your charitable giving'.[35] Similarly, in advance of the first Ethical Man broadcast, *Newsnight* presenter Gavin Esler asked 'What does it take to live ethically nowadays? What might it mean? A carbon neutral, environment friendly, green lifestyle? Or something different?'.[36] In practice, such questions went largely unanswered in both the Ethical Man broadcasts and Hickman's columns, and the focus was overwhelmingly on environmental issues. The fact that they are often asked is nevertheless significant, in that it appears to express a feeling that adopting a green lifestyle is not enough and that something more thoroughgoing and radical is needed.

Part of the problem, for those wishing to take a more radical and politicised stance, is that the mainstreaming of environmentalism and CSR has made ethical consumption by individuals or self-auditing by businesses unexceptional. Prime Minister Tony Blair asked UK companies to produce environmental reports by the end of 2001 (Litter 2009: 66), so the *Guardian* was not even a particularly early adopter of the increasingly widespread trend for ethical auditing.

> We can do something that's unique, different from just any other company. We can set an example, and we can reach our audiences. Our audience's carbon footprint is 10,000 times bigger than ours...
>
> That's the carbon footprint we want to conquer.
>
> We cannot do it with gimmicks. We need to reach them in a sustained way. To weave this issue into our content – make it dramatic, make it vivid, even sometimes make it fun. We want to inspire people to change their behavior.[37]

Although these sentiments would not have been out of place in one of the *Guardian's* ethical audits, they are from a 2007 speech by the chairman of News Corporation, Rupert Murdoch, announcing his intention to make his company's operations carbon neutral. The conservative stance of Murdoch's news outlets, such as Fox News in the US or the *Sun* in the UK, is very different from the *Guardian's* self-image as the leading voice of liberal and socially progressive opinion, yet News Corporation achieved its environmentally-friendly goal in 2011, and indeed its UK satellite TV operation had already become carbon neutral by the time Murdoch set the same goal for the entire company.

In 2003, the *Guardian* may have been a step ahead of Murdoch in self-auditing, but it was following in the wake of other large corporations whose moves in this direction it had reported scathingly the previous year. One of the paper's leading environmental writers, George Monbiot, had lambasted the whole idea of CSR and denounced the former UK director of Greenpeace, Lord Peter Melchett, for taking up a post with a PR firm to produce such reports (Monbiot 2002). Similarly, noting that 'McDonald's, Rio Tinto, Nike, Nestlé and British American Tobacco have all produced "sustainability reviews" or CSR reports', *Guardian* correspondent Felicity Lawrence found it 'impossible not to be cynical about this latest fashion', describing these corporations as 'rushing to get the latest business fashion accessory: a conscience'. Yet she did not entirely dismiss their efforts, arguing that 'the notion of CSR reporting represents a genuine effort to consider the wider role of business in civil society'. Reading between the lines, it seems that Lawrence differentiated a 'genuine effort' from mere 'greenwash' in terms of how far companies embraced a broader anti-consumerism:

> It is no coincidence that the list of companies leading the way with CSR reporting reads like a roll call of the anti-capitalists' pet hates....The vogue for social responsibility in business has been around for some time – wearing, in the main, environmental clothes. But just going green was so-o-o early 90s. The anti-capitalist protests in Seattle made companies realise that they would have to do more to preserve their reputations.
>
> (Lawrence 2002)

Perhaps surprisingly, given the *Guardian's* own commitment to environmentalism, merely 'going green' or wearing 'environmental clothes' is seen as superficial and insufficient. It was already so mainstream that something more radical was required in order to demonstrate real commitment.

As Barnett *et al.* (2005: 21) note, there is a distinction to be made between two senses of 'ethical consumption' (though in practice they are often intertwined): on the one hand, consumption is seen as 'a medium for moral and political action', whereby choosing the ethical, sustainable, or fair trade option expresses moral choices; on the other, there is also the idea that consuming 'ethically' means reducing overall levels of consumption, where consumerism itself is taken as the object of moral evaluation. Emphasising the more anti-consumerist aspects of ethical living is often understood as the way to turn it into a more radical and 'political' approach. Yet as we saw in Chapter 1, anti-consumerism has a long history in mainstream Anglo-American politics. In truth, the idea that anti-consumerism is some kind of radical, left-wing position was mistaken all along, as some writers on contemporary environmentalism have recognised. Leigh Phillips (2015: 27), for example, makes a robust case that a 'campaign against economic growth and overconsumption should have no place on the left'. Similarly, Matthias Varul (2013: 297) argues that anti-consumerism 'reneges on the idea of self-emancipation' and is instead a 'movement of an enlightened few trying to wean the intoxicated masses off their addiction to consumption'.

A further limitation is that those taking a more radical, anti-consumerist approach still tend to reproduce therapeutic, inward-focussed modes of agency. This is clear from other ethical living experiments featured in the paper. In July 2004, a few months after Hickman's column started in the *Guardian*'s Consumer section, Lucy Siegle launched a weekly 'eco column' in the *Guardian*'s sister Sunday publication, the *Observer*. She began by explaining that 'about five years ago my eco conscience began to form', due to the 'drip drip drip effect of images of melting polar icecaps, landfills spewing rubbish, and crops sprayed with pesticides'. Siegle soon decided, however, that simply limiting her own 'ecological footprint' was not enough:

> It would have ended there – with organic box-scheme delivery, eco light bulbs and a small composting bin – but then I felt the urge to live more ethically, adding a raft of social justice criteria to my environmental checklist. And so it was no longer just about minimising my eco footprint, but making sure I hadn't stamped on anyone else in the process....Instead of buying impulsively, I have to go through a sort of consumerist catechism. Who made this product? Why did they make it? Why do I need it?
>
> (Siegle 2004)

Taking aim at the 'McWorld culture' of 'global corporations' and 'international branding', Siegle argued that 'Being part of a consumer society increasingly means buying into a perpetual consume-and-chuck-it happy hour that is unsustainable'. Her proposed solution (to 'take an inventory of everyday occurrences and habits – from eating a Sunday roast to going to the gym and shopping on the high street – and give them an ethical work-out') still centred on consumption. But the more radical, political content was to be found less in the practices themselves than in the meaning given to them – the internal 'catechism' through which Siegle wrestled with her conscience – since this was an inward-directed project of working on the self. As Siegle reassured her readers: 'You don't need to be a heroic whistleblower or front-page activist when you can be a legend in your own living room'.

A few months after Siegle started her ethical living column, another *Guardian* writer, Neil Boorman, tried living without 'consuming anything supplied by a corporation for seven days' (Boorman 2004). Although Boorman's report on his one-week trial was tongue-in-cheek, lampooning 'woolly puritans' as much as 'retail zombies', in 2006 he undertook a more earnest, sustained and dramatic experiment which culminated in him publicly burning all his branded possessions. He maintained a blog recounting the progress of his personal de-branding, which he then turned into a book, excerpts of which were published in the *Guardian* in 2007. His speech at the finale event – the 'Bonfire of the Brands' – sounded a radically anti-consumerist note:

> I am a member of a generation that has been sold to from the day it was born. And what you see today is the result...a person who is addicted to brands...I've been buying these things, believing they would make me more successful, more likable, more sexy....These brands are nothing but an expensive con. In the UK, we are £200bn in debt over this stuff.
>
> (Boorman 2007)

Boorman was disappointed to find that burning his possessions did not entirely free him from consumer culture: a magazine placed him 'among 2007's key trend drivers in a minority swing towards Nu Austerity, a resurging conscience culture that will facilitate a rise of thrift chic' (Boorman 2007). He subsequently went on to work for a marketing and advertising agency. The key word above, though, is 'addicted'. Boorman repeatedly explained the rationale for his life makeover in these terms, as wanting to 'rid myself of an addiction to these brands' and to 'find happiness somewhere other than in a department store':

> Gradually it began to dawn on me that the value system by which I judged myself and those around me was hollow. I lived a comfortable life with my partner. Both my parents were alive and remained happily married. I had a wide circle of friends, I had an interesting career and, it goes without saying, I possessed a lot of nice, branded things. I should have been happy. Instead, I felt empty, cheated and disillusioned.
>
> (Boorman 2007)

He seems to have understood his behaviour as literally 'addictive', much in the way that other authors at the time treated consumerism as a mental health disorder (for example, James 2007, Naish 2008). Whether intended literally or metaphorically, there could be no clearer articulation of the therapeutic framing of 'ethical' living than the notion of breaking an addiction to consumerism in order to find true happiness.

In the run-up to the 2009 Copenhagen climate summit, the *Guardian* coordinated a joint editorial, published simultaneously by 56 papers in 45 countries, in many cases on the front page. As deputy editor Ian Katz explained: 'the global leader carries a simple message to the politicians and negotiators gathered in Copenhagen: if all of us, who disagree about so much, can agree on what must be done, then surely you can too' (Katz 2009). It exemplified the sort of consensual media coverage that, as we saw in Chapter 2, Max Boykoff and many others hoped would promote greater action on the environment (Boykoff and Boykoff 2007), but also demonstrated its limitations. As Chris Methmann observes:

> The tremendous failure of the Copenhagen climate change summit in December 2009 condenses all the paradoxes of present day global climate politics in one political event. Never before had so much attention been directed at global warming in the media, politics and society....Never before had a bigger summit dealt with climate change. Never before had one single political event been charged with so much hope. And never before had the disappointment been worse.
>
> (Methmann 2014: 1)

Ironically, it seems that the critique of the post-political climate change consensus was given a boost by the waning of that consensus. It was not just the disappointing outcome of the summit itself: a combination of circumstances shook the unanimity that had been established in the mid-2000s and made it increasingly clear that the consensus-building approach had not worked as expected. Just before the summit came the 'climategate' scandal about leaked emails exchanged between scientists at the University of East Anglia's Climatic Research Unit which, sceptics claimed, showed manipulation of evidence on climate change (Monbiot 2009). This then coloured coverage of Copenhagen (see Gavin and Marshall 2011; Terracina-Hartman and Oshita 2013), and was followed in early 2010 by the admission that the IPCC's 2007 report had included an unsupported claim that the Himalayan glaciers would disappear by 2035 (Carrington 2010).

Perhaps the most important factor, however, was the on-going economic crisis which followed the 2008 financial crash. Previously, environmentalist advocates had sometimes argued in favour of economic recession. The *Guardian's* George Monbiot, for example, urged people to 'riot for austerity';[38] argued for 'a campaign not for abundance but for austerity' (2006: 215); and maintained that 'only a recession makes sense' (2007a). Once there actually was a recession, such views no longer sounded so radical. The *Guardian's* Copenhagen editorial still argued that 'kicking our carbon habit' would require us to 'change our lifestyles' and to 'pay more for our energy', but urging consumer restraint and recommending higher energy bills now sounded more like an apology for government austerity policies rather than an ethical rejection of rampant consumerism.[39] At the end of 2007, Monbiot acknowledged the limitations of 'demand-side policy for tackling climate change':

> Most of the governments of the rich world now exhort their citizens to use less carbon. They encourage us to change our lightbulbs, insulate our lofts, turn our TVs off at the wall....Without supply-side policies, runaway climate change is inevitable, however hard we try to cut demand.
>
> (Monbiot 2007b)

By 'supply-side policies', he meant that the campaigning focus should shift instead to 'leaving fossil fuels in the ground'. The *Guardian* took up this theme in 2015, launching a major new campaign in the run-up to the Paris climate change summit, under the slogan 'Keep it in the ground'. Editor Alan Rusbridger presented it as a personal legacy project, set in motion as he approached retirement, to address his one regret from his time in charge of the paper: that 'we had not done justice to this huge, overshadowing, overwhelming issue' (Rusbridger 2015a). The new campaign was to be both pragmatic and principled, arguing the case for disinvestment from fossil fuel companies on both financial and moral grounds, and while it was clearly targeted at finance directors and investment fund managers (with a particular focus on lobbying the Wellcome Trust and the Bill and Melinda Gates Foundation), aiming to 'force the issue now into the boardrooms and inboxes of people who have billions of dollars at their disposal' (Rusbridger 2015b), it also aimed to involve readers. The campaign was launched in partnership with the environmental NGO 350.org ('building a global movement') and signed up over 226,000 supporters from more than 170 countries within the first six months (Randerson 2015).

The *Guardian* evidently felt that the campaign was a daring and dramatic departure, drawing its readers' attention to 'the risks of telling the biggest story in the world' (Howard 2015a). Yet in many respects the results seemed similar to past practice. The campaign was underpinned not so much by politics as by physics: the cause would 'almost certainly be won in time', said Rusbridger (2015b), because 'the physics is unarguable'. Despite the 'supply-side' focus, the campaign still involved advising readers on personal self-auditing, for example by offering guidance on 'How to divest your personal finance' (Howard 2015b). *Guardian* writers continued to debate the 'small steps' that individuals could take to address such an 'overwhelming' problem (LaBrecque 2015); and still focused on the problem of 'carbon addiction', urging readers to 'go green even if it makes you feel a pious twit' (Rustin 2015). It also turned out that 'the biggest risk of all' in telling the 'biggest story in the world' was that the Guardian Media Group might not divest its own funds from fossil fuels (Howard 2015a). The company's decision, in April 2015, to begin 'a process of divesting its more than £800m investment portfolio from fossil fuels' was presented as a major victory for readers (Rusbridger 2015c), although many readers may have been rather surprised that the paper still had millions of pounds invested 'unethically' after so many years of ethical self-auditing.

Most significantly, although Monbiot was doubtless correct to point out that he was 'the first person to suggest in the media that the best means of addressing climate change is to leave fossil fuels in the ground' (Monbiot 2015), there was a seven year delay between him suggesting it and his own paper taking notice. The more direct impetus for the *Guardian*'s campaigning stance in 2015 seems to have come from elsewhere: in 2014, President Barack Obama, World Bank Group President Jim Yong Kim, Bank of England governor Mark Carney, and former Goldman Sachs CEO and Secretary of the US Treasury Hank Paulson all highlighted the risks of continued investment in fossil fuel industries (Howard and Carrington 2015). This looks more like a case of following the lead of elite sources than taking some bravely independent and oppositional political stance.

Conclusion: the limits of lifestyle

That the *Guardian* would have taken its lead from elite figures should come as no surprise in light of the way that, as we saw in Chapter 2, news coverage of climate change has consistently followed the contours of elite debate. Critics have rightly argued that research on the media and climate change has been mistaken in assuming the need to press for more depoliticised, consensual news coverage. However, it would not be quite right to suggest that the media themselves are somehow 'to blame' for depoliticisation, as if the problem was simply one of journalistic choices. As suggested above in discussing the BBC's Ethical Man, it is more that the mainstreaming of climate change and the high degree of elite consensus allows and encourages the presentation of the issue in the way that Berglez and Olausson (2014) characterise as ideological: as an unquestioned, agreed truth to be incorporated into personal life-narratives and everyday practices. Moreover, as we saw in relation to the *Guardian*'s more partisan advocacy stance on climate change, attempts to make such mainstream discourses more radical by accentuating anti-consumerist ideas have stayed within the bounds of a therapeutic framing of the issue, including at the institutional level, whereby self-auditing and self-monitoring practices suggest a mode of 'political' action that tends to be introspective and inward-directed. Ultimately, what is at issue in the post-political is not so much whether media framing or news discourse has particular depoliticising effects, but the larger question of how political agency has been reinterpreted and redefined.

The issue raised by Anthony Downs in 1972, of whether environmental problems could be addressed via relatively straightforward policy changes and technological solutions or whether they necessarily required much more fundamental social and economic change, is still relevant to contemporary debates. Yet much has changed since the early 1970s. In the run-up to the 2009 Copenhagen summit, *Newsnight* broadcast a special edition putting the 'Greens on Trial'. Justin Rowlatt contributed two film reports and presenter Emily Maitlis then put points from the reports to representatives of the contemporary green movement.[40] Rowlatt's first film asked whether environmentalism was 'too conservative', because it was unwilling to embrace high-tech solutions such as nuclear power, genetic modification and nanotechnology; the second asked if greens were 'too radical', in adopting an anti-capitalist, anti-consumerist stance. To illustrate the latter, Rowlatt used a clip from the radical campaigning film *The Age of Stupid*, and Maitlis then asked the film's director, Franny Armstrong, whether the solution to climate change was 'basically smashing capitalism'. Armstrong's response was that the solution was for 'everybody, whether that's individuals, businesses, the military, the monarchy, to commit to cutting their emissions by 10%...in 2010'. This was the objective of the 10:10 campaign group that Armstrong had founded, and which, she claimed, had already attracted support from '10,000 businesses'. No doubt Armstrong was reluctant to be pigeonholed as an extremist, but it was nevertheless telling that she was keen not only to emphasise the mainstream nature of her campaign, but also to explain that its answer to the challenge that 'all of life on earth is at stake' was essentially that people should 'change their lifestyles a little bit': 'it's driving your car a bit less, changing your diet a bit, flying a bit less, lagging your loft'. The modernist rationalities implicit in either technological fixes or top-down policy solutions no longer seem credible to many people, not just to environmentalists: the precautionary attitude to risk is not only a feature of green thought, but holds sway across many areas of mainstream policy-making (Beck 1992, Runciman 2004). Most climate change campaigners seek much broader social change rather than technological solutions which leave the basic character of industrial society untouched, and many do indeed associate this with an anti-capitalist outlook. Yet in the absence of a modernist conception of politics, 'anti-capitalism' tends to be translated into individual lifestyle changes.

Notes

1   Ethical Man's water retention: http://news.bbc.co.uk/1/hi/programmes/newsnight/5210772.stm; Ethical Man's wind problems: http://news.bbc.co.uk/1/hi/programmes/newsnight/8351237.stm.

2   Go green or else: Ethical Man's story, 4 March 2007, http://news.bbc.co.uk/1/hi/programmes/panorama/6412547.stm.

3   *The Real Good Life* was cancelled after its first two episodes. Although *No Waste Like Home* completed its run, attempts to repeat the formula have generally not worked, particularly in the more didactic variant. In the 2007 BBC series *Outrageous Wasters* a team of 'eco-experts' sought to 'transform Britain's most wasteful households'. While their homes were given a 'radical eco-makeover, designed to shock them into changing their ways', the families were sent to an 'eco boot camp' called the 'House of Correction', where they were 'forced to face up to the impact of their wasteful ways and learn to live a carbon neutral lifestyle'. The show ran on BBC Three for only four episodes. Lucy Siegle, whose ethical lifestyle journalism for the *Observer* is discussed below, was to have been the presenter of a similar BBC Two show in 2008, *Guilt Trip*, which reportedly 'challenged "pampered people" to "make the ultimate sacrifice for the good of the planet"', but it was cancelled while still in production (McNally 2008). One of the most successful examples was the BBC Two series *It's Not Easy Being Green*. In 2006 it followed the efforts of Dick Strawbridge and his family to live a green lifestyle as they moved to a dilapidated farmhouse; and, in a second series in 2007, to help others 'go green and conserve the environment'. It returned for a third series in 2009, but as more of a magazine-style programme.

4   Diary of an Ethical Man, 28 March 2006, http://news.bbc.co.uk/1/hi/programmes/newsnight/4758834.stm; An Ethical Blog, 9 August 2006, http://www.bbc.co.uk/blogs/newsnight/2006/08/an_ethical_blog_1.html.

5   Ethical Man begins challenge (video), 22 February 2006, http://news.bbc.co.uk/1/hi/programmes/newsnight/8351236.stm.

6 I am the ethical man, 22 February 2006, http://news.bbc.co.uk/1/hi/programmes/newsnight/4736228.stm.

7 Ethical Man (video), 13 February 2009, http://news.bbc.co.uk/1/hi/programmes/newsnight/7888971.stm.

8 See the BBC's editorial guidelines on 'maintaining impartiality': http://www.bbc.co.uk/guidelines/editorialguidelines/advice/conflicts/2maintainingimp.shtml.

9 Ethical Man – home (video), 28 March 2006, http://news.bbc.co.uk/1/hi/programmes/newsnight/7906227.stm.

10 Ethical Man goes vegan (video), 3 June 2008 (first broadcast February 2007): http://news.bbc.co.uk/1/hi/programmes/newsnight/7433753.stm.

11 The BBC was embroiled in a row about its 'impartiality' over climate change in 2007, following its decision to cancel a special 'Planet Relief' programme (see Plunkett 2007, Shanahan 2007).

12 A transcript of the programme is available at http://news.bbc.co.uk/1/hi/programmes/panorama/6428165.stm. The film can be watched at http://www.veoh.com/watch/v391651spkXPyHr/climatechange.

13 See: http://resolve.sustainablelifestyles.ac.uk/about-resolve. Jackson's work might more accurately have been characterised as advocacy for sustainable consumption (see for example Jackson 2004).

14 Ethical Man begins challenge (video), 22 February 2006, http://news.bbc.co.uk/1/hi/programmes/newsnight/8351236.stm.

15 Ethical man's ethical sister, 24 Feb 2006, http://news.bbc.co.uk/1/hi/programmes/newsnight/4745102.stm.

16 See, for example, Hickman's reports for the *Guardian* on: Meat and dairy (15 July 2004), Domestic energy (23 September 2004), and Clothing (7 October 2004).

17 Ethical Man begins challenge (video), 22 February 2006, http://news.bbc.co.uk/1/hi/programmes/newsnight/8351236.stm.

18 Ethical Man – home (video), 28 March 2006, http://news.bbc.co.uk/1/hi/programmes/newsnight/7906227.stm.

19 A year without a car (video), 15 July 2008 (first broadcast March 2006), http://news.bbc.co.uk/1/hi/programmes/newsnight/7508352.stm.

20 A flight that almost cost me my marriage (video), 4 December 2009 (first broadcast October 2006), www.bbc.co.uk/blogs/ethicalman/2009/12/the_tv_stunt_that_almost_cost_me_my_marriage.html. See also 'Ethical Man my a***', 17 October 2006, http://www.bbc.co.uk/blogs/newsnight/2006/10/ethical_man_my_arse.html.

21 Being Mrs Ethical, 28 March 2006, http://news.bbc.co.uk/1/hi/programmes/newsnight/4852722.stm; Ethical Wife in cash giveaway shock, 20 February 2007, www.bbc.co.uk/blogs/newsnight/2007/02/ethical_wife_in_cash_giveaway_shock.html.

22 Being Mrs Ethical, 28 March 2006, http://news.bbc.co.uk/1/hi/programmes/newsnight/4852722.stm.

23 See their exchange in the comments section at: Forage with care, 13 Nov 2006, www.bbc.co.uk/blogs/newsnight/2006/11/forage_with_care.html; and Bee Rowlatt's article on her reinvestment of her husband's shares: Ethical Wife in cash giveaway shock, 20 February 2007, www.bbc.co.uk/blogs/newsnight/2007/02/ethical_wife_in_cash_giveaway_shock.html.

24 One of Rowlatt's online commentaries exactly matched Berglez and Olausson's illustration of this narrativisation and its wistful or nostalgic – but typically light – emotional tone: 'When I was a child we could make snowmen. That isn't possible now' (2014: 59). In Rowlatt's case, he found old cine-camera footage from his childhood showing him and his sisters making a snowman: 'Do you remember snow?', he asked, 'It is beginning to look like my kids will be lucky to ever build a snowman in our garden' (Who remembers winter?, 12 Jan 2007, www.bbc.co.uk/blogs/newsnight/2007/01/who_remembers_winter_1.html). Fittingly, in 2010 *Newsnight* chose Rowlatt to do a report on that winter's exceptionally heavy snow: How does snow make you feel?, 6 January 2010, http://news.bbc.co.uk/1/hi/programmes/newsnight/8443808.stm.

25 Kitchen experiment 'proves' science of global warming, 17 December 2009 (video excerpt from *Newsnight's* 16 December edition), http://news.bbc.co.uk/1/hi/programmes/newsnight/8417920.stm. After witnessing the experiment, an audience member who had initially confessed his scepticism about global warming, said: 'I'd be an idiot if I didn't agree. I totally now believe it's a fact. I totally believe that I'm responsible, we're responsible.'

26 A global warming experiment in my kitchen, 16 December 2009, http://news.bbc.co.uk/1/hi/programmes/newsnight/8417295.stm. For a critique of the experiment, see: Anthony Watts, BBC botches grade school CO_2 science experiment on live TV – with independent lab results to prove it, *Watts Up With That*, 24 December 2009, https://wattsupwiththat.com/2009/12/24/bbc-botches-grade-school-co2-science-experiment-on-live-tv-with-indepedent-lab-results-to-prove-it/.

27 We are all ethical men and women now, 13 April 2007, www.bbc.co.uk/blogs/newsnight/2007/04/we_are_all_ethical_men_and_women_now.html.

28 *Living Our Values* 2004, p41, extracted online as 'Readers care and want to read about the environment', www.theguardian.com/environment/2004/sep/17/ethicalliving.lifeandhealth.

29 See: Auditor's statement 2003–04, www.theguardian.com/values/socialaudit/story/0,,1303554,00.html. The point was reinforced in the 2005 edition which reported that Hickman's 'ethical living audit of his own life' had subsequently been 'developed into a weekly ethical living section' (*Living Our Values* 2005, p34, http://image.guardian.co.uk/sys-files/Guardian/documents/2005/10/28/Environment.pdf).

30 *Living Our Values* 2003, p27, http://image.guardian.co.uk/sys-files/Guardian/documents/2003/12/04/environment.pdf. In a striking example of 'behind-the-scenes' influence, Natalie Bennett joined the UK Green Party in 2006, worked as deputy editor and then editor of the *Guardian Weekly* from 2007 until 2012, and then became leader of the Green Party from 2012 to 2016.

31 *Living Our Values* 2003, p27, http://image.guardian.co.uk/sys-files/Guardian/documents/2003/12/04/environment.pdf.

32 See 'Theory into practice' in the Environment section of the 2004 report, extracted online as 'What difference does writing about the environment make?', www.theguardian.com/values/socialaudit/environment/story/0,,1305106,00.html.

33 Hickman drew the same parallel in the 2004 report, p41, extracted online as 'Readers care and want to read about the environment', www.theguardian.com/environment/2004/sep/17/ethicalliving.lifeandhealth.

34 *Living Our Values* 2003, p33: 'All cleaning products used in GNL offices are biodegradable except for its toilet cleaner which contains phosphoric acid in order to remove limescale'.

35 Your ethical feedback (5), 5 Jun 2006, http://news.bbc.co.uk/1/hi/programmes/newsnight/5049432.stm; Your ethical feedback (1), 27 February 2006, http://news.bbc.co.uk/1/hi/programmes/newsnight/4748934.stm; Ethical advice, 28 February 2006, http://news.bbc.co.uk/1/hi/programmes/newsnight/4756792.stm.

36 Details of tonight's programme, 22 February 2006, http://news.bbc.co.uk/1/hi/programmes/newsnight/4740718.stm.

37 Global Energy Initiative Launch: Rupert Murdoch's Remarks, 9 May 2007, www.theaustralian.com.au/news/rupert-murdochs-speech-on-carbon-neutrality/news-story/a0aec5b8ae16beccad083dbf0cb22768.

38 Quoted in Mason 2004. Monbiot inspired an activist group, Riot4austerity (https://green365.wordpress.com/about-the-riot4austerity), whose main focus appears to have been their own consumer habits and household energy consumption.

39 Fourteen days to seal history's judgment on this generation, *Guardian*, 7 December 2009, www.theguardian.com/commentisfree/2009/dec/06/copenhagen-editorial.

40 Greens on Trial, *Newsnight*, BBC Two, 14 October 2009. See: Greens on trial: Are they too radical? (http://news.bbc.co.uk/1/hi/programmes/newsnight/8308957.stm); Greens speak out on claims they are too radical (http://news.bbc.co.uk/1/hi/programmes/

newsnight/8308989.stm); Greens on trial: Are they too conservative? (http://news.bbc.
co.uk/1/hi/programmes/newsnight/8308923.stm); and Greens speak out on claims they
are too conservative (http://news.bbc.co.uk/1/hi/programmes/newsnight/8308943.stm).

References

Barclay, Liz and Michael Grosvenor (2007) *Green Living for Dummies*. London: Wiley.

Barnett, Clive, Philip Cafaro and Terry Newholm (2005) Philosophy and Ethical
Consumption, in Rob Harrison, Terry Newholm and Deirdre Shaw (eds.) *The Ethical
Consumer*. London: Sage.

Barnett, Clive, Paul Cloke, Nick Clarke and Alice Malpass (2011) *Globalizing Responsibility*.
Chichester: Wiley.

Beck, Ulrich (1992) *Risk Society: Towards a New Modernity*. London: Sage.

Berglez, Peter and Ulrika Olausson (2014) The post-political condition of climate change:
An ideology approach, *Capitalism Nature Socialism*, 25 (1): 54–71.

Boorman, Neil (2004) Branded for life, *Guardian*, 23 October, www.theguardian.com/
film/2004/oct/23/culture.features.

Boorman, Neil (2007) Name dropper, *Guardian*, 25 August, www.theguardian.com/
lifeandstyle/2007/aug/25/shopping.features.

Boykoff, Maxwell T. and Jules M. Boykoff (2007) Climate change and journalistic norms: A
case-study of US mass-media coverage, *Geoforum*, 38: 1190–1204.

Carrington, Damian (2010) IPCC officials admit mistake over melting Himalayan
glaciers, *Guardian*, 20 January, www.theguardian.com/environment/2010/jan/20/ipcc-
himalayan-glaciers-mistake.

Carvalho, Anabela and Tarla Rai Peterson (2012) Reinventing the Political, in Anabela
Carvalho and Tarla Rai Peterson (eds.) *Climate Change Politics: Communication and Public
Engagement*. Amherst, NY: Cambria Press.

Craig, Geoffrey (2016) Political participation and pleasure in green lifestyle journalism,
Environmental Communication, 10 (1): 122–41.

DeLaure, Marilyn (2011) Environmental comedy: No impact man and the performance of
green identity, *Environmental Communication*, 5 (4): 447–66.

Downs, Anthony (1972) Up and down with ecology – the 'issue-attention cycle', *The Public
Interest*, 28: 38–50.

Gavin, Neil and Tom Marshall (2011) Mediated climate change in Britain: Scepticism on
the web and on television around Copenhagen, *Global Environmental Change*, 21 (3):
1035–44.

Gow McDilda, Diane (2007) *Everything You Need to Know about Green Living*. Cincinnati,
OH: David and Charles.

Hailes, Julia (2007) *The New Green Consumer Guide*. London: Simon and Schuster.

Hickman, Leo (2004) Ethical living challenge: How green is your house? *Guardian*, 29
January, www.theguardian.com/environment/2004/jan/29/foodanddrink.ethicalmoney.

Hickman, Leo (2006) *A Life Stripped Bare: My Year Trying to Live Ethically*. London: Eden
Project Books in association with the *Guardian*.

Hickman, Leo (2008) *A Good Life: The Guide to Ethical Living*. London: Eden Project Books
in association with the *Guardian*.

Hickman, Leo (2014) A year of ethical living revisited, *Guardian*, 11 March, www.theguardian.
com/lifeandstyle/2014/mar/11/a-year-of-ethical-living-revisited.

Howard, Emma (2015a) What are the risks of telling the biggest story in the world?, *Guardian*, 9
April, www.theguardian.com/environment/keep-it-in-the-ground-blog/2015/apr/09/
what-risks-telling-biggest-story-in-world-climate-change-podcast.

Howard, Emma (2015b) How to divest your personal finance – your questions answered, *Guardian*, 26 May, www.theguardian.com/environment/keep-it-in-the-ground-blog/live/2015/may/26/keep-it-in-the-ground-qa-how-to-divest-your-personal-finances.

Howard, Emma and Damian Carrington (2015) Everything you wanted to ask about the *Guardian's* climate change campaign, *Guardian*, 16 March, www.theguardian.com/environment/2015/mar/16/everything-you-wanted-to-ask-about-the-guardians-climate-change-campaign.

Illouz, Eva (2007) *Cold Intimacies: The Making of Emotional Capitalism*. Cambridge: Polity.

Isin, Engin F. (2004) The neurotic citizen, *Citizenship Studies*, 8 (3): 217–35.

Jackson, Tim (2004) Models of Mammon: A Cross-Disciplinary Survey in Pursuit of The 'Sustainable Consumer' (Working Paper Series Number 2004/1). Guildford: University of Surrey Centre for Environmental Strategy, http://citeseerx.ist.psu.edu/viewdoc/download?doi=10.1.1.136.2607&rep=rep1&type=pdf.

James, Oliver (2007) *Affluenza*. London: Vermillion.

Katz, Ian (2009) How the climate change global leader project came about, *Guardian*, 6 December.

LaBrecque, Sarah (2015) Climate change is overwhelming – so should we focus on small steps?, *Guardian*, 25 March, www.theguardian.com/sustainable-business/behavioural-insights/2015/mar/25/climate-change-is-overwhelming-small-steps.

Lawrence, Felicity (2002) Social butterflies, *Guardian*, 19 August, www.theguardian.com/world/2002/aug/19/globalisation.voluntarysector.

Lewis, Tania (2008) Transforming citizens? Green politics and ethical consumption on lifestyle television, *Continuum*, 22 (2): 227–40.

Littler, Jo (2009) *Radical Consumption: Shopping for Change in Contemporary Culture*. Maidenhead: Open University Press.

Mason, Paul (2004) Looking to the hydrogen horizon, BBC, 21 September, http://news.bbc.co.uk/1/hi/sci/tech/3675760.stm.

McNally, Paul (2008) BBC axes 'guilt' TV show over exploitation fears, *BrandRepublic*, 30 April, www.brandrepublic.com/article/806110/bbc-axes-guilt-tv-show-exploitation-fears.

Methmann, Chris (2014) *We Are All Green Now: Hegemony, Governmentality and Fantasy in the Global Climate Polity* (PhD diss.). Hamburg: Staats- und Universitätsbibliothek Hamburg, http://d-nb.info/1050818407/34.

Monbiot, George (2002) Greens get eaten, *Guardian*, 15 January [available at www.monbiot.com/2002/01/15/greens-get-eaten/].

Monbiot, George (2006) *Heat: How We Can Stop the Planet Burning*. London: Allen Lane.

Monbiot, George (2007a) In this age of diamond saucepans, only a recession makes sense, *Guardian*, 9 October 2007 [available as 'Bring on the recession: How else will the destructive effects of growth be stopped?', at www.monbiot.com/2007/10/09/bring-on-the-recession].

Monbiot, George (2007b) Leave it in the ground, *Guardian*, 11 December 2007 [available at www.monbiot.com/2007/12/11/rigged/].

Monbiot, George (2009) Pretending the climate email leak isn't a crisis won't make it go away, *Guardian*, 25 November, www.theguardian.com/environment/georgemonbiot/2009/nov/25/monbiot-climate-leak-crisis-response.

Monbiot, George (2015) Why leaving fossil fuels in the ground is good for everyone, *Guardian*, 7 January, www.theguardian.com/environment/georgemonbiot/2015/jan/07/why-leaving-fossil-fuels-in-ground-good-for-everyone.

Naish, John (2008) *Enough: Breaking Free from the World of More*. London: Hodder and Stoughton.

Ouellette, Laurie and James Hay (2008) Makeover television, governmentality and the good citizen, *Continuum*, 22 (4): 471–84.

Paterson, Matthew and Johannes Stripple (2010) My Space: Governing individuals' carbon emissions, *Environment and Planning D*, 28 (2): 341–62.

Pepermans, Yves and Pieter Maeseele (2014) Democratic debate and mediated discourses on climate change: From consensus to de/politicization, *Environmental Communication*, 8 (2): 216–32.

Phillips, Leigh (2015) *Austerity Ecology and the Collapse-Porn Addicts*. London: Zero Books.

Plunkett, John (2007) BBC drops climate change special, *Guardian*, 5 September, www.theguardian.com/media/2007/sep/05/bbc.television2.

Randerson, James (2015) A story of hope: The *Guardian* launches phase II of its climate change campaign, *Guardian*, 5 October, www.theguardian.com/environment/2015/oct/05/a-story-of-hope-the-guardian-launches-phase-two-of-its-climate-change-campaign.

Rose, Nikolas (1999) *Powers of Freedom*. Cambridge: Cambridge University Press.

Runciman, David (2004) The precautionary principle, *London Review of Books*, 26 (7): 12–14, www.lrb.co.uk/v26/n07/david-runciman/the-precautionary-principle.

Rusbridger, Alan (2015a) Climate change: Why the *Guardian* is putting threat to Earth front and centre, *Guardian*, 6 March, www.theguardian.com/environment/2015/mar/06/climate-change-guardian-threat-to-earth-alan-rusbridger.

Rusbridger, Alan (2015b) The argument for divesting from fossil fuels is becoming overwhelming, *Guardian*, 16 March, www.theguardian.com/environment/2015/mar/16/argument-divesting-fossil-fuels-overwhelming-climate-change.

Rusbridger, Alan (2015c) The Guardian Media Group's decision to divest is in thanks to our readers, *Guardian*, 2 April, www.theguardian.com/environment/keep-it-in-the-ground-blog/2015/apr/02/the-guardian-media-groups-decision-divest-thanks-to-our-readers.

Rustin, Susanna (2015) Go green. Even if it makes you feel a pious twit, *Guardian*, 15 December, www.theguardian.com/commentisfree/2015/dec/15/go-green-carbon-footprint-climate-change.

Schneider, Jen and Glen Miller (2011) The impact of 'no impact man': Alternative hedonism as environmental appeal, *Environmental Communication*, 5 (4): 467–84.

Shanahan, Mike (2007) Getting the balance wrong, *Guardian*, 5 September, www.theguardian.com/commentisfree/2007/sep/05/gettingthebalancewrong.

Siegle, Lucy (2004) Ethical living: Clean and serene, *Observer Magazine*, 25 July, www.theguardian.com/lifeandstyle/2004/jul/25/shopping.ethicalliving.

Silver, James (2007) The accidental green hero, *Guardian*, 5 March, www.theguardian.com/media/2007/mar/05/mondaymediasection8.

Terracina-Hartman, Carol and Tsuyoshi Oshita (2013) Climate change on trial: An analysis of the media coverage of climategate, *International Journal of Climate Change*, 4 (3): 119–32.

Thomas, Lyn (2008) Alternative realities: Downshifting narratives in contemporary lifestyle television, *Cultural Studies*, 22 (5): 680–99.

Varul, Matthias Zick (2013) Towards a consumerist critique of capitalism: A socialist defence of consumer culture, *Ephemera*, 13 (2): 293–315.

Varul, Matthias Zick and Dana Wilson-Kovacs (2008) *Fair Trade Consumerism as an Everyday Ethical Practice – A Comparative Perspective*, ESRC/University of Exeter, http://people.exeter.ac.uk/mzv201/FT%20Results.pdf.

4
CLIMATE CHANGE AND CELEBRITY CULTURE

Several authors have been struck by the 'celebritisation' of climate change in the early twenty-first century. Support for global charities is 'practically part of the contemporary celebrity job description and a hallmark of the established star' (Littler 2008: 238–9). 'Being green has become trendy in Hollywood' (Thrall *et al.* 2008: 370). The 'environment – and in particular, climate change – is the new black' (Goodman forthcoming: 1). Celebrity environmental activism is not a new phenomenon (Brockington 2009: 13; Meyer and Gamson 1995: 181), but many commentators have identified a significant increase in the scope and scale of celebrity environmental campaigning in recent years. It has been suggested that celebrities have become 'more and more elevated as authoritative and seemingly "authentic" voices' on the issue (Anderson 2013: 340), and that they have 'positioned themselves as increasingly powerful and politicised meditators' of our relationship with the environment (Goodman forthcoming: 1). By 2010 the association between the environment and celebrity culture was already so close that Max Boykoff *et al.* could wryly suggest that celebrities were starting to supplant polar bears in the public imagination, as 'the new "charismatic megafauna" for climate awareness, understanding and engagement' (Boykoff *et al.* 2010: 2; Boykoff and Goodman 2009: 399).

Other causes have also attracted much celebrity attention in recent years, of course – the other particularly prominent areas of activity being humanitarianism and international development (see, for example, Brockington 2014, Chouliaraki 2013, Kapoor 2013). Indeed, celebrity interest in climate change has to be seen as part of a larger change, sometimes characterised in terms of a coming together of celebrity culture and charity – 'charitainment' (Thrall 2008: 377) or 'celanthropy' (Rojek 2014) – but also understood in terms of what Philip Drake and Michael Higgins (2006: 87) call 'the increasingly interwoven nature of celebrity and politics'.[1] This chapter focuses on this interweaving of celebrity and politics, as a symptom of the post-political.

In assessments of celebrity activism there is often an implicit assumption that there is still a clearly delineated arena of politics proper, in relation to which celebrity campaigning could operate, for good or ill, effectively or unsuccessfully (Hammond 2009). Yet the sphere of public political activity has also been transformed by its intermingling with celebrity culture. This chapter follows the distinction that John Street (2004: 437–8) makes between two types of 'celebrity politician': celebrities who speak out on public and political issues, and politicians who behave like and/ or associate with celebrities. In relation to celebrity climate campaigning, Leonardo DiCaprio and Emma Thompson are discussed below as examples of *activist celebrities* – people who have used their fame from the world of entertainment as a platform to campaign around the issue of climate change. Here and in the next chapter we also look at *celebrity activists* – individuals who initially came to prominence through more conventional political/campaigning activity but who have latterly achieved a kind of celebrity status – focusing on the examples of Al Gore and Naomi Klein. Gore has worked extensively with musicians and other celebrities on *Live Earth* and related projects, earning him the image of a 'mega rock-star carbon warrior' (Ervine 2013: 100); while Klein has come to be seen as a celebrity public intellectual (Goodman forthcoming: 1), whose 2014 book about climate change merited an author profile in the style magazine *Vogue* (Powers 2014).

These examples have been chosen so as to try to give some sense of the range of different activities and outlooks that exist: Thompson's work with Greenpeace follows a 'traditional' model of celebrity endorsement, lending a famous face to an NGO campaign, while DiCaprio is more like an independent non-state actor, producing campaigning films and running his own charitable foundation; Gore's background in mainstream politics informs what some have criticised as a too-conservative outlook on climate change which leaves existing power structures intact, while Klein purports to offer a more radical view aiming at root-and-branch social change. While these contrasts are important, however, there are also common threads running through different versions of 'celebrified' climate change, converging in a shared therapeutic discourse.

Celebrity standing

The obvious difficulty in combining environmental concern with celebrity culture is the potentially jarring incompatibility between a generally anti-consumerist outlook which urges us to 'reduce, reuse, recycle', and the conspicuous luxury consumption associated with the entertainment industries. Commenting on *Vanity Fair*'s second annual 'Green Issue' in 2007, for example, Max Boykoff and Mike Goodman (2009: 395) express their astonishment at how the 'magazine of record for Hollywood haute couture politics and culture' could, without irony, place articles highlighting environmental crisis alongside advertisements for 'meteorically-expensive designer clothing, watches, and sunglasses'. Similarly, the *Vogue* profile of Naomi Klein mentioned above enthused that despite her anti-capitalist views she is not 'earnest, smugly righteous, [and] out of touch', but rather is 'unobtrusively stylish' and 'understands the joys of shopping' (Powers 2014).

Such incongruities may affect the credibility of celebrities to speak about their chosen causes, particularly climate change. Boykoff and Goodman (2009: 400), for example, highlight Leonardo DiCaprio's defensive comments about his air travel at the 2007 Cannes Film Festival. DiCaprio was in Cannes promoting his climate change documentary, *The 11th Hour*, but was also defending Al Gore, who had been criticised for his 'extensive travel' to promote his 2006 film, *An Inconvenient Truth*. 'The way he travels and the way he lives his life should not be criticised' said DiCaprio, perhaps hoping that the same principle might extend to himself. Reporting DiCaprio's assurances that he tries whenever possible to take commercial flights, the *Guardian*'s Charlotte Higgins (2007) remarked tartly that 'while a small but increasing number of Britons are giving up flying altogether, for the Hollywood A-lister there is a preliminary step: eschewing private jets'. Accusations of hypocrisy because of his use of yachts and private planes are regularly levelled at DiCaprio (Cronin 2015, Mohr and Smith 2016); while Gore's huge personal energy use – his Nashville mansion alone consumes around 20 times more than the US average – has also been a focus of frequent criticism (Ervine 2013: 102). Wealthy celebrities are an easy target for such criticisms, which circulate both as a subgenre of celebrity gossip stories, and sometimes also as part of deliberate efforts to throw doubt on the claims that celebrity supporters make for environmental causes.[2]

Not only are celebrities' lifestyles liable to be scrutinised for evidence of their commitment to the campaigns they espouse, but their knowledge of the issues involved is also often called into question. In publicising his 2015 film *The Revenant*, for example, DiCaprio remarked on the difficulty of filming winter scenes in Canada because of the warm weather, interpreting this as evidence of climate change. He was mocked in the Canadian media for not understanding that the warm air was a 'Chinook', a common local weather phenomenon (Duncan 2015), and then mocked again after alluding to it once more in his February 2016 Oscars speech accepting the award for Best Actor for the film. As one report put it, 'Environmentalists worry [the] actor's use of [the] common chinook as [a] case in point for global warming damages [his] credibility' (Robson 2016). Similarly, a claim by Emma Thompson in an interview with the BBC's *Newsnight* programme that global temperatures were set to 'rise four degrees Celsius by 2030' attracted criticism when it was mentioned in an August 2016 report by the Corporation's governing body as an instance where an inaccurate claim should have been challenged by BBC journalists.[3] Although the report – about the BBC's handling of statistics – dealt with many other issues and examples, and despite the fact that her interview had been broadcast almost a year earlier, media headlines tended to focus on Thompson's inaccurate claim (for example, Furness 2016). Again the effect was to undermine not only the celebrity, but also the cause for which she spoke.

These sorts of criticisms point up the importance of what David Meyer and Joshua Gamson (1995: 189) call 'standing' – the 'socially constructed legitimacy to engage publicly in a particular issue'. Since the fame of most celebrities derives, at least initially, from activities outside the world of politics and public affairs, when they intervene in that world the basis of their authority to do so needs somehow to be established: celebrities represent no specific social interest or constituency and are 'typically expected to speak as individuals' (1995: 202). The status of celebrities, Meyer and Gamson note, is 'detached from institutional centers, from requirements of merit or achievement, and from solid grounds of authenticity and authority', and is instead 'marked by tremendous ambiguity and instability' (1995: 184). It might be thought that the media treatment of celebrities who speak about climate change, as in the examples above, is sometimes unfair and perhaps motivated in part by antipathy to the cause they represent. In Meyer and Gamson's analysis, however, it is the very nature of media-manufactured celebrity culture that is the underlying problem. The mediated construction of celebrities gives them a kind of 'semi-fictional' status, at once both 'real and artificial, spontaneous and programmed, performing themselves and being themselves', so that their 'standing as authentic, and their motives for public action...are always in question' (1995: 184). This emphasis on standing and authenticity is useful for understanding the emotional connections that campaigning celebrities attempt to make, and we shall return to it further below. It also relates, however, to another characteristic of celebrity campaigning highlighted by Meyer and Gamson: a preference for consensus.

Celebrity involvement in a campaign, Meyer and Gamson suggest, brings a pressure to 'depoliticize or deradicalize' its claims, thereby 'institutionalizing and domesticating dissent' (1995: 188). This aspect of celebrity activism might seem especially relevant to our concern with the post-political, and certainly in Meyer and Gamson's example – a late-1980s/early-1990s campaign to protect Walden Woods in Massachusetts from planned housing development – the involvement of celebrities does indeed seem to have been the key factor in pushing the campaign toward a more consensual, moderate and non-conflictual stance. Yet it is questionable how significant this is in the case of climate change campaigning, which is already widely understood in terms of what Meyer and Gamson describe as 'consensus-style politics and broad coalition building' (1995: 191), rather than this being a result of celebrity involvement. More germane for our discussion is their point that celebrities tend to 'gravitate to less challenging movements', preferring '"liberal" causes such as civil rights, environmentalism, and homelessness' (1995: 188). This is because it is easier to 'claim legitimate standing' in relation to issues that are less controversial and which have a broad relevance, such as the environment, where 'celebrities, regardless of status and wealth, can easily claim shared concerns and standing with a larger public' (1995: 190). Celebrity involvement with climate change is an indication of its already-established status as a consensual, mainstream concern, although one might see the preference for consensus as working against more radical, anti-capitalist versions of environmentalism.

Indeed, increased celebrity attention to climate change can be understood as part of the high point of consensus on the issue which, as discussed in Chapter 2, was established by the mid-2000s – a period when, as Boykoff *et al.* (2010: 3) show, news coverage of celebrities and climate change rose markedly, reaching a peak in 2007. The website *Look to the Stars*, for example, which aims to 'publicize the many wonderful things that celebrities are doing to help the world',[4] was established in 2006 – the year that *An Inconvenient Truth* was released, and the year that entertainment and fashion magazine *Vanity Fair* published its first 'Green Issue' (featuring Julia Roberts, George Clooney, Robert F. Kennedy Jr. and Gore on the front cover).[5] The following year, when DiCaprio's climate change documentary *The 11th Hour* was released, Gore won an Oscar for *An Inconvenient Truth*, won the Nobel Peace Prize (shared with the IPCC) for his campaigning on climate change, and staged the massive *Live Earth* 24-hour spectacular, featuring nine concerts across seven continents, broadcast live via television, radio and online, attracting an estimated audience of two billion (Anderson 2011: 541).

This heightened celebrity environmentalist activity could be understood in similar terms to the way that Ilan Kapoor (2013: 37, 115) characterises celebrity humanitarianism: as 'pundits of global altruism', he argues, celebrities fit into and strengthen the contemporary 'postdemocratic order' by helping to 'construct the social fantasy of a humanized and "caring" capitalism'. Somewhat similarly, Boykoff *et al.* (2010: 9–10) argue that the efforts of the 'environmental "celebritariat"' are severely limited if not downright counter-productive, in that they 'ramify contemporary forms of consumerism' and 'reduce proposed critical behavioral changes to the domain of barely effective individual fashion-fad rather than influence substantive long-term shifts in popular discourse and political action'. They have in mind things like the efforts of celebrities such as Daryl Hannah, who 'has the accoutrement-completing, green celebrity-required website devoted to "sustainable" lifestyle advice and shopping' (Boykoff *et al.* 2010: 4); or the UK NGO Global Cool (motto: 'Making Green Mainstream'), launched in 2007 with a reception at 10 Downing Street hosted by Tony Blair and attended by actors and musicians.[6] Boykoff and Goodman (2009: 404) sardonically sum up the problem as the promotion of 'conspicuous redemption': celebrities are represented as 'heroic individuals' engaged in 'extraordinary personal action' to 'save the planet' (via mundane things like 'changing one's light bulb or buying a hybrid automobile'), thereby focusing on individualised 'solutions'. They worry that celebrity campaigning may well be effective, but in promoting the wrong approach: diverting attention away from the 'systemic and large-scale political, economic, social and cultural shifts that will likely be necessary to address the multifarious problems and difficult choices associated with modern global climate change' (Boykoff and Goodman 2009: 404).

We will return in Chapter 5 to the question of just how effective celebrity campaigning is, but we first need to clarify the nature of the appeal that celebrities are making. Rejecting any 'simplistic understanding of celebrities as functioning only in terms of "role models"' (Boykoff *et al.* 2010: 6), these critics turn instead to P. David Marshall's view of celebrities as embodying 'a discursive battleground on the norms of individuality and personality within a culture', and as offering 'subject positions that audiences can adopt or adapt in their formation of social identities' (Marshall 1997: 65, quoted in Boykoff and Goodman 2009: 398). Yet if virtue-signalling celebrities do indeed offer a redemptive, 'glorified and righteous path to decarbonisation', it seems unlikely to work simply through the media promotion of 'high-profile "heroes" that cements individualization and the neoliberal project'

(Boykoff and Goodman 2009: 404). As we argued in Chapter 3, the 'responsibilised' subject of contemporary modes of (neoliberal) governance is not a 'heroic', robust and autonomous individual but an emotional, self-monitoring, 'neurotic citizen' (Isin 2004: 232).

Far from presenting themselves as 'heroic' saviours, celebrities often seem to adopt a pointedly self-deprecating demeanour as environmental advocates. A striking example is DiCaprio's 2016 climate documentary, *Before the Flood*. As news clips play in the background criticising and undermining him by deriding his lack of credibility and expertise, DiCaprio acknowledges his ignorance: 'The truth is, the more I've learned about this issue and everything that contributes to the problem, the more I realise how much I don't know, how much I don't know about this issue'. Partly, of course, this sets him up as a proxy for the viewer so that as he embarks on a journey to 'see for myself what's going on and what can be done', he takes us with him. Yet it is not just that he lacks knowledge: his tone at the start of the film is distinctly downbeat and doubtful. 'I want to do everything I can to learn more about this issue', DiCaprio tells us, 'but it all kinda seems beyond our control'. Later in the film, as we watch him walking through an airport and getting into a car, he also admits: 'my [carbon] footprint is probably a lot bigger than most people's, and there are times when I question, what is the right thing to do?' DiCaprio even confesses that he is unsure whether he can fulfil his role as United Nations Messenger of Peace on climate change: 'If the UN really knew how I feel, how pessimistic I am about our future, I mean, to be honest, they may have picked the wrong guy'. Again one might see this as a narrative device characteristic of 'reality' genres – introducing an element of jeopardy about whether he will be able to meet the challenge – but it does mean that the tone is very far from heroic.

Similarly, Naomi Klein's opening line in her 2015 film *This Changes Everything* is: 'Can I be honest with you? I've always kinda hated films about climate change. What is it about those vanishing glaciers and desperate polar bears that makes me want to click away?' Like DiCaprio, she positions herself, and by implication the viewer, as not really engaged or convinced that she can make a difference. In her accompanying book, Klein (2014: 3) writes that: 'I denied climate change for longer than I care to admit. I knew it was happening, sure....But I stayed pretty hazy on the details and only skimmed most of the news stories'. Like DiCaprio, she also admits that she lives a 'high consumer [lifestyle]' and has a 'shiny card in my wallet attesting to my "elite" frequent flyer status' (2014: 2–3). At least in the cases examined here, there is little sign of celebrities setting themselves up as 'heroic individuals' (Goodman 2013: 10).[7]

This may seem like an inconsequential point, but it is important because critics of celebrity activism make a similar move to the one that we encountered in Chapter 3 regarding discussions of 'ethical' or 'green' consumption. As we saw, critics are usually unwilling to write off consumer-led approaches entirely, and indeed despite their scathing comments about 'individualised and "commodified" solutions', Boykoff *et al.* also maintain that, 'one can viably argue that celebrity endeavours possess the potential to enrol a new set of "actors" by mobilizing audiences and fan bases that otherwise would not find interest in climate change' (2010: 5, 10).There is an attempt to disaggregate the negative, individualistic, acquisitive consumerism from the positive, responsible, connected and reflexive ethics. So on one hand Boykoff *et al.* argue that approaches which invite an 'individualised, responsibilised response' are 'taking a cue from the playbook of neo-liberalism' in locating 'climate change problems and solutions at the level of the individual'; yet on the other hand they maintain that by appealing to 'the inspired individual consumer-citizen' celebrity actions may help to 'pioneer a new-millennium reflexivity, usher in meaningful change and inspire emergent engagements and movements' (2010: 9–10). Similarly, Mike Goodman (2013) wrestles back and forth with the positives and negatives of 'celebritus politicus'. On the one hand, celebrities are 'consummate commodities', whose promotion of causes is simultaneously a promotion of their own 'brand', and who as 'heroic individuals' have the 'power to market to us the "right" pathways and products to deal with global inequalities and environmental concerns' – pathways and products that promote 'commodified and monetized forms of care politics' and thereby entrench the 'moral authority of neo-liberalised sustainabilities' (2013: 75, 81–2). On the other hand, though, celebrities can also enact a 'non-commodified, "fan-collectivising" function', whereby the audiences who are attracted to 'enviro-tainment' can then be mobilised and their influence brought to bear 'through the bodies, foundations and access of *Celebritus politicus* to the halls of power' (Goodman 2013: 74).

In much the same way that analysts of ethical consumption wish to discard the individualist consumerism but accentuate the ethics, so critics of celebrity campaigning would like to reject the commodified 'heroic individualism' but retain the apparent influence of celebrity campaigners for good causes. The fact that there is not really much evidence of the 'heroic' celebrity image they want to criticise, however, suggests that possibly the critique is missing the target and misunderstanding the nature of the appeal being made by campaigning celebrities. In their more recent work, Goodman, Boykoff and others have suggested that the potentially positive aspects of celebrity activism are to be found specifically in its emotional, affective dimension, and it is to this that we turn next.

Emotional pedagogy and 'After Data' climate communication

While acknowledging the 'power of mediated spectacle to distort and de-politicize', Goodman *et al.* (2016: 680) wonder whether it may also 'contain conditions for more radical critique', asking: 'How can spectacular environmentalisms be deployed and do work for the powerless and in the support of rights and justice?' Their answer, in essence, is that it can do this to the extent that it has an emotional dimension: 'spectacular environmentalisms…are not simply about the transmission of facts, words and "rational" knowledge but also about fostering emotion and ecologies of feeling' (2016: 681). The celebrity-driven initiatives of this 'spectacular' style of environmental campaigning can 'get our attention and pique our environmental imaginaries in ways that work to get us to feel, to connect and to "do"', offering 'pedagogical narratives about how we should go about caring for more-than-human nature' and 'affective, para-social performances of anger, sadness, loss, hope, joy and many other emotions that attempt to frame our own affective responses to save the world' (2016: 681).

Similarly, Julie Doyle *et al.* (forthcoming: 14–15) emphasise the need for climate change communication to include an 'emotional/affective' dimension. They argue that celebrities' ability to 'embody and generate a range of feelings and affects about climate change' means that they can 'reach out to audiences' in a way that more distancing, rational or scientific communication strategies may not. Although Doyle *et al.* (forthcoming: 2–3) reiterate the concern that celebrity advocacy 'may render climate change as a commodity media spectacle', they argue that celebrities may also 'politicize emotions that remain circumscribed by neoliberal solutions and action' and thereby help to 'move us beyond scientific data and facilitate more emotional and visceral connections with climate change'.

These authors usefully highlight the importance of emotion in contemporary celebrity campaigning. Yet while their accounts are descriptively accurate in many ways, it is argued here that the emotional appeal of celebrity climate advocacy should be understood differently. First, there has not been quite as dramatic and 'profound' a change as they suggest (Doyle *et al.* forthcoming: 20); and second, celebrity campaigning involves, not the politicisation of emotion, but the increasing personalisation and emotionalisation of environmental advocacy. What is happening is not that emotion is being added on as an extra layer or dimension to make the communication of a message more effective, but that the style and tone of celebrity 'spectacular environmentalisms' are shaped by the specific conventions of celebrity media genres and by the broader cultural context, both of which tend to foreground the personal and emotional. Instead of a completely new direction, what we see is the unfolding of the post-political logic of celebrity climate change campaigning, towards a therapeutic rather than a political discourse.

Doyle *et al.* (forthcoming: 19–20) argue that celebrities are at the forefront of a marked shift in climate communication, from 'dry accounts of the latest scientific knowledge' to 'stories of personal and/or literal journeys'. They draw a pointed contrast between DiCaprio's *Before the Flood* and the sorts of films that were being produced a decade earlier: Gore's *An Inconvenient Truth* (2006) and DiCaprio's *The 11th Hour* (2007). Whereas these earlier films focused on 'data, information and knowledge', teaching the public about climate change in 'minute PowerPoint detail', now we have entered an 'After Data era of climate communications', in which information and argument have given way to 'accentuated celebrity emotion and affect' (forthcoming: 24–5). Celebrity climate advocacy appears promising even to otherwise sceptical critics because it is 'not simply about the transmission of facts, words and "rational" knowledge but also about fostering emotion and ecologies of feeling' (Goodman *et al.* 2016: 681).

This hearts-and-minds approach exactly describes the premise of the 2014 and 2016 Showtime/National Geographic TV series *Years of Living Dangerously*, which set out to combine 'the blockbuster storytelling styles of Hollywood's top movie makers' with 'the investigative skills of...leading national news journalists and scientists'. The programmes featured a mixture of celebrities and high-profile journalists taking on the role of 'correspondent', investigating, but also emotionally reacting to, the impacts of climate change. As one of the show's celebrity participants, actor/musician Jack Black, explains:

> With regard to climate change there is the scientific facts and then there's the heart and the emotion. If you only have one without the other they're not really effective. You need a combination of the two to really spark the flames of positive social change.[8]

Both for the programme-makers and for the critics, the idea seems to be that an ineffective information-deficit model of environmental communication needs to be supplemented with an emotion-deficit model (Norgaard 2011).

Yet, as suggested in the Introduction, there has been a periodic 'rediscovery' of the importance of emotion as the extra magic ingredient that might have the potential to make climate communication more effective. While Doyle *et al.* see *An Inconvenient Truth* as dry-as-dust and data-heavy, earlier critics viewed it as 'powerful and emotional' (Beattie *et al.* 2011: 111). In Geoffrey Beattie *et al.*'s study of viewer responses to the film, *An Inconvenient Truth* is seen as important because it goes beyond the limitations of 'so-called rational thought' (which they understand as often 'little more than a post-hoc justification for our behaviour'), and instead addresses the need for 'something much more concrete and much more emotional to persuade us that global warming is both real and urgent' (Beattie *et al.* 2011: 109, 107). The reiteration of these arguments suggests a longer-term development rather than a recent new departure.

Moreover, in relation to the other prominent areas of celebrity advocacy – international development and humanitarianism – critics have identified a similar (and again, earlier) shift toward personalisation and emotion, but with less positive outcomes. Lilie Chouliaraki compares the appeals made by Audrey Hepburn (United Nations Children's Fund Goodwill Ambassador from 1988–93) and Angelina Jolie (United Nations High Commissioner for Refugees Ambassador from 2001), finding 'significant variation in the authentication strategies of celebrities' over time (2012: 5; and see also Chouliaraki 2013, chapter 4). In contrast to Hepburn's 'controlled' and 'dispassionate' ambassadorial style, which focussed on drawing attention to the suffering of others, Jolie's engagement is 'constructed as part of a trajectory towards personal self-fulfilment...a moment of self-enlightenment', a rewarding project of the self which 'may intensify connectivity with the celebrity but does not necessarily facilitate a move to action' (2012: 11, 16). The intimate, emotional, confessional style of Jolie's celebrity witnessing may be unsuccessful in its overt aims since it risks focussing attention on herself rather than the cause, but it is in tune, Chouliaraki suggests, with 'a pervasive orientation of contemporary public discourse towards practices of self-disclosure as the performance of authenticity' (2012: 15).

Similarly, Jo Littler (2008: 238) describes how the 'confessional and intimate' character of Jolie's persona presents her charity work as 'a component of a narrative about the actress's "real" and "intimate" life', part of 'her story', her 'journey so far'. Both authors follow Luc Boltanksi (1999) in identifying this sort of emotionalised celebrity advocacy with an inadequate and limited 'politics of pity', as opposed to an approach focussed on global justice (although, as Littler (2008: 248) notes, it may sometimes 'mobilise the language of justice'). For Chouliaraki, 'confessional' celebrity witnessing limits responses to humanitarian suffering to the extent that it 'encourages a narcissistic disposition of voyeuristic altruism' (2012: 17). Littler (2008: 248) takes this argument a step further in suggesting that 'the construction of the celebrity as having a caring, compassionate soul might…be read as a means to compensate for and legitimate weakness: a weakness in a wider system'. The 'performance of celebrity soul, or the performance of the internalisation of social anguish', she argues, is 'an attempt to gesturally redress the insecurities of the system it is part of'; a way of acknowledging the suffering caused by global inequalities while not actually transforming them (Littler 2008: 248).

Claims about the positive potential of celebrities' emotional engagement with environmentalism may be sustained by an element of wishful thinking. As Goodman admits, 'we (or at least I!) have a deep-seated hope that our contemporary politicised glitterati actually *really* do care about the climate' (Goodman 2013: 8, original emphasis). Yet the sincerity of celebrities is not the issue. Instead, we need to examine what Chouliaraki (2012: 4) calls the 'production of authenticity, understood as the performative strategies by which [the] celebrity renders her impersonations of altruism credible and legitimate'. In referring to celebrity 'impersonations of altruism' or 'performances' of sincerity, she does not mean that these are necessarily false or disingenuous: rather, the point is that instead of 'didactic exhortations' celebrities exemplify and enact an 'altruistic disposition' as a positive value which the audience are presumed to share and aspire to (2012: 3). If the celebrity is, as Chouliaraki (2012: 2) suggests, 'a figure who commands the necessary symbolic capital to articulate personal dispositions of acting and feeling as exemplary public dispositions at given historical moments', we need to understand how and why these 'exemplary' emotional dispositions allow celebrities to act as 'morally-tinged, affective pedagogues' (Doyle *et al.* forthcoming: 25).

In other words, regardless of whether particular celebrities really mean what they say, the question is why emotional authenticity has become so central to their standing or credibility as spokespeople on the issues they take up. The answer, it is suggested below, is twofold: first, celebrity campaigning follows formats and conventions which are part of current media and celebrity genres; and second, in a context of declining political authority, celebrities' enactment of personal and emotional authenticity fills a gap of political legitimacy, not with a politicisation of emotion, but with a therapeutic sense of transcendence.

Emotional standing and celebrity genres

To test the proposition that 'affective climate media and celebrities work to specifically *politicize* emotion and affect in the context of climate change' (Doyle *et al.* forthcoming: 27, original emphasis), let us begin by re-examining the development toward an 'after data' mode of celebrity campaigning. There has indeed been a shift of emphasis in how celebrities establish their standing on the issue of climate change, from scientific expertise to emotional involvement.

DiCaprio's *The 11th Hour* is a conventional 'talking heads' documentary in which different experts speak to the camera, and is straightforwardly 'designed to teach the public about climate change' (Doyle *et al.* forthcoming: 19). DiCaprio provides some voiceover narration and is occasionally seen on screen as a link between the film's different segments: his main role is simply to be a familiar front-man who might be expected to attract viewers through his fame as an actor. He does not claim any personal standing on the issue, although he does invoke the mainstreaming of environmentalism in the manner that, as we saw earlier, Meyer and Gamson (1995: 190) suggest. 'Environmentalism was once the project of a passionate few' DiCaprio tells us, leaving open the question of whether he was one of the few, but now 'environmentalism has become today a broader unifying human issue'. At the beginning of the film he says that 'we've reached out to independent experts on the frontlines of what could be the greatest challenge of our time', and the documentary's authority derives from the standing of these experts. A decade later in *Before the Flood*, in contrast, experts take second place to the depiction of 'a significant personal journey for DiCaprio' in which he appears 'front and center as our serious, earnest and caring, emotive and affective guide and male "lead"' (Doyle *et al.* forthcoming: 20–1).

Yet at the time, the creators of *The 11th Hour* understood what they were doing in terms of making an emotional connection with the audience rather than simply delivering information to them. The film's co-writer/director, Nadia Conners, drew a contrast in this respect with *An Inconvenient Truth*, which she saw as 'a very fact-based film while *The 11th Hour* is more of an emotional experience about our place in the world'.[9] As we have seen, *An Inconvenient Truth* in turn was also understood as having a powerful emotional dimension. In particular, it includes a number of passages which bear only tangentially on the substantive issues tackled in the film but relate instead to Gore's nostalgic account of his own personal experiences (see further Murray and Heumann 2007). Of course *An Inconvenient Truth* does contain a lot of detailed information about climate change, and Gore establishes his standing on the issue in part by mentioning scientific experts and other public figures who are 'namedropped' as personal friends and acquaintances.[10] Yet his standing is also established in emotional terms through the autobiographical passages.

In Chapter 1 we saw how, when running for Vice-President in 1992, Gore framed climate change as a moral rather than a political issue and invoked his own personal history as a basis for claiming moral authority and a sense of purpose. The same thing happens in *An Inconvenient Truth* as Gore recounts his political and, especially, personal losses and setbacks. He explains how the near-tragedy of his son's car accident made him re-prioritise the issue of climate change. Similarly, there is a sequence telling the story of Gore's sister, Nancy, who died of lung cancer, prompting their father to give up farming tobacco. This relates only obliquely to the main issue (the point being that people are sometimes slow to make changes even when faced with evidence), but it is highly emotive, not only because of the subject matter of family bereavement but also because of the formal qualities of its presentation. This portion of the film is illustrated with black-and-white family photographs, black-and-white documentary footage of a young woman on a cigarette production-line, and film of Gore revisiting the family farm, with some of the latter sequence having been made to appear as if it is old home-movie footage shot with a cine-camera.

The visual evocation of Gore's childhood memories here recalls a similar sequence earlier in the film in which his reminiscences about a childhood spent partly on his father's farm are illustrated with old newspaper cuttings, and again with black-and-white family photographs and cine-camera-style footage. This earlier sequence, in turn, also repeats in a different form the opening shots of the film, which feature footage of a riverbank (presumably at or near the farm) to accompany Gore's softly spoken voiceover:

> You look at that river, gently rolling by. You notice the leaves rustling in the wind. You hear the birds, you hear the tree frogs. In the distance you hear a cow. You feel the grass; the mud gives a little bit on the riverbank. It's quiet, peaceful. And all of a sudden, it's a gear shift inside you. And it's like taking a deep breath and going [sigh], 'oh yeah, I forgot about this'.

Gore thus establishes a relationship between an idealised natural environment and personal change and remembrance. This evocative opening is later recalled visually in a photo of the family boating on the river, and in the repetition of the riverbank footage, now with the cine-camera effect added.

Taken together, the effect of these sections of the film is to evoke a powerful sense of nostalgia for his own childhood as a time of harmony and closeness with nature, but one which becomes interconnected, via the personal stories of the (near) death of family members, with a sense of grief. These intimate and emotional passages help to personalise the argument and to lend it credibility in the sense of relating it to his own authentic experience. In a sense, then, Gore's film is an early example, perhaps even a template – its data-heavy PowerPoint detail notwithstanding – for the conventional format of 'personal journey' celebrity engagements with climate change that critics have described in later texts.

The US premiere of DiCaprio's *Before the Flood* was screened on National Geographic Channel immediately after the first episode of Season Two of *Years of Living Dangerously* (on 30 October 2016). Audiences would have seen a near-seamless continuity in format: *Years* also features charismatic celebrities (and a few celebrity journalists) going on witnessing journeys to observe, learn about and react to the effects of climate change, just like DiCaprio does in *Before the Flood*. Since both are mainstream Hollywood products, the generic resemblance is not particularly surprising. But the genre conventions of how to depict a genuine celebrity experience and convey authentic emotional reactions are broader than this: similarities can be found across different issues and different national contexts, sometimes popping up in unexpected places.

As illustration, consider Emma Thompson's contribution to Greenpeace's successful *Save The Arctic* campaign, as part of which she and her 14-year-old daughter, Gaia Wise, travelled to the Arctic in 2014. As well as featuring in Greenpeace's own publicity, she worked with the *Guardian* newspaper to make a short film of the trip.[11] At the start of the *Guardian* film, Thompson declares: 'I think if you're informed enough, and your experience is a genuine experience you can be helpful'. She has often emphasised the general need to become informed about climate change,[12] but here she makes very little reference to information and expertise. There is no explicit mention of global warming until three-quarters of the way in, and even then it is alluded to only briefly.[13] Thompson visits an Arctic research centre where we see her in the audience attentively taking notes at a talk by an Ecotoxicologist, yet we do not get to hear the talk, and her to-camera reaction is about his style of delivery more than the content ('I love the fact he told us all these terrible things with enormous, rather cheerful enthusiasm'). Instead, the main point of her involvement with the campaign was clearly to be seen having a 'genuine experience'. As she reflected later: 'I spoke to climatologists and scientists there, and I understood in a very visceral, real way, standing there on mountain glaciers, what really was happening at the top of the world and its impact on not only local communities, but on the entire world'.[14]

An analysis of Danish 'celebrity narratives of global caring' for development causes by Lene Bull Christiansen and Birgitta Frello (2016: 134) describes a mode of 'emotional address' which closely corresponds to Thompson's Arctic film, despite the different issue and the different national production context. Narratives of celebrity witnessing, they observe, typically show the celebrity travelling 'in a way that involves giving up the amenities and ease of Western wealth and privilege' and showing that s/he is 'willing to let go of comforts, and be shown in less than flattering ways'; displaying 'some form of genuine emotion and personal attachment' in order to establish authenticity; and demonstrating that s/he has 'experienced personal growth on an emotional level' (Christiansen and Frello 2016: 137–43). In the *Guardian* film, Thompson light-heartedly complains that her cold-weather gear makes her 'look like a sofa', and we see her join in with the chores on board the ship, helping with the cleaning while Joanna Kerr, executive director of Greenpeace Canada, jokes that at home 'Emma loves her toilet so much she keeps her two Oscar statuettes in there'. While such scenes show the authenticity of the experience in the sense of Thompson roughing it, emotional authenticity is communicated in facial expressions, tone of voice and vocalisation of feelings. For example, Thompson describes the reactions we can observe her having on screen, saying things such as: 'When I got to the ice that really blew me away'. Emotional growth or impact is mainly located in the effects of the expedition on the teenage Gaia. 'My daughter is a very brave, she's a courageous and intelligent girl, and I knew that she would be very much affected by this', explains Thompson. Similarly, towards the end of the film, she describes how: 'It broke us leaving that ship. For a 14 year old this is a life shaper. It's something that she'll never forget, because I think in many ways she's found her tribe'. This mention of 'tribal' identification, though, is as close as we get to anything even remotely approaching a 'politicisation' of emotion.

Perhaps critics understand the depiction of celebrities' emotional reactions to climate change as potentially politicising simply because they would like to read it that way. Looked at differently, celebrity witnessing journeys and displays of emotional authenticity have a lot in common with non-campaigning media formats made purely for entertainment. Certainly the 'celebrity journey' has become a very familiar TV genre: in 2010 the *Guardian's* TV review was already complaining that it was overused, asking 'How much longer can this format of "another celebrity, another self-indulgent journey" be flogged?'[15] In April 2016, Emma Thompson staged an anti-fracking protest, also for Greenpeace, in which she and her sister, fellow actor Sophie Thompson, re-enacted a popular British TV programme, *The Great British Bake Off*, even though baking was completely irrelevant to the issue at hand.[16] As well as providing a media-friendly, entertaining format for the protest, it also afforded a framework for the 'behind-the-scenes' displays of emotional engagement and the mild dramatic tension created by elements of competition and jeopardy characteristic of this style of programming.[17]

The generic features of celebrity campaigning can be performed in different ways: Thompson's style is quite light and whimsical, while others have greater intensity and seriousness. Yet even a show like *Years of Living Dangerously*, with its dramatic tone and more journalistic mode of address, works partly because it feeds the voyeuristic appetites of celebrity culture, offering intimate glimpses of its stars. In the programmes themselves, we see the celebrities 'off-duty', as it were, wearing ordinary clothes, being themselves; and the programme's website gives viewers abundant bonus material in the form of short 'Web Exclusives' taking us 'Behind the Scenes', or offering personal testimonies from individual celebrities explaining 'Why they are involved' or 'Why they care'.[18] Arnold Schwarzenegger explains that having been for many years on a 'fitness crusade', his experience as Governor of California led him to believe that he had to 'fight this new crusade, the environmental crusade', for example. Jack Black explains, in terms reminiscent of Chouliaraki's and Littler's descriptions of humanitarian celebrities, that he has been on an 'incredible personal journey'. Others invoke a 'personal connection' (America Ferrera) or affirm that they are 'passionate about it' (Don Cheadle). We could interpret these testimonies as teaching us how to feel about climate change, but we could also see them as reminiscent of the sort of things that contestants say on reality shows like *Bake Off* or *MasterChef* (both of which have 'celebrity' variants), when asked to speak about their motivation. The genre requires participants to vocalise their commitment by saying things like 'I'm passionate about cooking', or 'I'm passionate about climate change', as appropriate. The articulation of such authentic, personal, emotional reactions, in other words, is not in itself evidence that there is anything political going on.

Therapeutic transcendence

An alternative way to understand the political meaning of celebrity emotional display is suggested by Littler's argument that the therapeutic, confessional style of contemporary celebrity advocacy meets a broader ideological need: 'plugging the gap' of legitimacy, as she puts it (Littler 2008: 248). These ideas about the ideological role of displays of 'celebrity soul' are echoed by Dan Brockington (2008: 553–4), who suggests that 'the prominence of celebrity in environmentalism's affairs is a necessary consequence of the alienations of capitalism'. The proposition that celebrity environmentalism provides a consolation for an experience of alienation entails a two-step argument: first, it is suggested that our mediated, 'para-social' relationship with celebrities fills a void created by an everyday experience of social isolation, by providing an experience of 'intimacy' with people we do not know (2008: 562). This applies to celebrity generally, rather than celebrity campaigning around the environment or other issues, and although it does fit nicely with the confessional and intimate character of contemporary celebrity humanitarianism as described by Chouliaraki and others (Brockington and Henson 2014: 5), it fits equally well with accounts of celebrity 'emotion work' in other, non-campaigning, contexts (Nunn and Biressi 2010: 49–50). Second, specifically in relation to the environment, it is argued that since this alienation is caused by 'urban living' it also involves an 'alienation from nature', so that when people consume celebrity environmentalism 'they are restoring their relationship with the wild' (Brockington 2008: 558, 563). Conservation is the main focus of Brockington's (2008, 2009) work on celebrity environmentalism, and in that particular context perhaps an idea of closeness to nature may be part of the appeal, but it seems less obviously applicable to climate change campaigning.

Like those writers who criticise celebrity activism as only superficially addressing symptoms rather than engaging with the deeper causes of problems, Brockington draws a contrast between a genuine conservationism that is 'well grounded in local places', and 'safe' celebrity support for 'remote...exotic places', 'general abstract causes' and 'generic ideals', which he understands as an inauthentic and 'ungrounded' environmentalism (2008: 553–4, 559–60). Climate change seems a better fit for the more generic and general category. Attachment to place tends to involve 'remote' or 'exotic' locations, as Brockington suggests. In the *Guardian* film discussed above, for example, Emma Thompson says that 'The idea of there being a sanctuary that people can visit and see its beauty and bring back that in their hearts is a wonderful, wonderful dream to have'. The Arctic features as a magical place where one can have a special emotional experience.

Other celebrity journeys, such as DiCaprio's in *Before the Flood* or many of those featured in *Years of Living Dangerously*, include meetings with people affected by climate change, whose experience of the environment is 'grounded' in particular places in the way that Brockington thinks is authentic and meaningful. Yet the celebrities – and by extension, the viewing audience – are necessarily temporary visitors, observing this authentic experience with a 'tourist gaze' (Urry 2002). It is notable that in *Before the Flood* DiCaprio tends to relate the things he sees on his travels to his own main area of experience: acting. Reflecting that the impacts of climate change will mean that 'everything that we now take for granted' will be different in the future, he says: 'I feel like I'm in some weird, surreal movie'. As he flies over the Canadian tar sands in a helicopter, he remarks that it looks like Mordor from the film *The Lord of the Rings*. Recalling the time when he first heard about climate change (from Al Gore), he remembers it sounding 'like some nightmarish science fiction film'. And as he considers what can be done, he says that 'if this was a movie we could write the ending of this script, and we could figure a way out of this mess'. There is indeed an authenticity to these perceptions, but it is rooted in his experience of Hollywood cinema.

Doyle *et al.* describe how *Before the Flood* 'accentuates and showcases emotions and affects':

> the smiles and sincerity of Elon Musk who is ready to deploy his battery business and entrepreneurial skills in service of a carbon-free future, the dire warnings of Ban Ki-Moon and, of course, those of the main witnessing muse of DiCaprio who marvels at the 'violence' of icebergs calving into the ocean, the surprise of being confronted about his own personal climate impacts and his hopeful tone in discussions of easy climate 'wins'.
>
> (Doyle *et al.* forthcoming: 21)

While these emotional tones are certainly present, however, this reading is limited in that the affective power of the film does not really come from a succession of individual moments. More to the point is their observation that the film depicts a 'significant personal journey for DiCaprio' and that it is 'shot through with stories of his early childhood' (forthcoming: 20–1). In fact there is just one story of his early childhood in the film, but it is indeed the key to understanding the film's emotional journey.

There are two overall narrative arcs to *Before the Flood*: one following DiCaprio's public role as a UN Messenger of Peace; the other involving a more personal story of his spiritual, indeed religious, reflection on climate change. Both involve emotional encounters as described by Doyle *et al.*, but the emotional heart of the film is the second story: this private, 'personal journey' informs and gives emotional meaning to the first, public narrative, but it does not 'politicise' it. Rather, it personalises the issue in the sense that it shows us the significance of climate change for DiCaprio in very personal terms.

The ostensible story, the overt narrative of the film, begins with DiCaprio's appointment as a Messenger of Peace at the September 2014 UN climate summit, where he makes a public speech but also, as noted above, in voiceover expresses private doubts about his suitability for the role and his lack of knowledge. He then embarks on a literal journey, travelling to distant places and meeting many different people, to find out more. At the end of the film, he returns to the UN, this time addressing the delegates at the April 2016 signing of the Paris climate agreement and urging them to take action. In this final speech, he mentions his journey and as he names the different places we have seen with him, there are brief cuts back to earlier scenes of those locations. The implication is that the first-hand experience gained through the journey has now settled his doubts, supplied the missing knowledge, and supports an unambiguous call for decisive action. As he puts it in the speech, in terms reminiscent of Ingolfur Blühdorn's (2007: 253) account of the contemporary 'performance of seriousness', discussed in the Introduction: 'After 21 years of debates and conferences, it is time to declare no more talk, no more excuses, no more ten-year studies'.

Although the relevant passage is not shown in the film, DiCaprio had actually expressed very similar views in his September 2014 UN speech, emphasising the need for 'decisive large scale action' and saying 'This is not a partisan debate...not a question of politics'.[19] So we might see the initial self-doubt as no more than a narrative device to introduce an element of jeopardy and dramatic tension, or we might see it as the genuine expression of feelings that he did not air publicly at the time but did share 'privately' with viewers of the film, who see behind the scenes of the public façade. Either way, the film encourages the sense that doubt has been resolved and uncertainty replaced with conviction in the course of the narrative journey. The overriding message is one of consensus agreement – there is no need for 'more talk': all that is needed now is action. The emotionality of the delivery could certainly be said to convey urgency and conviction, and thereby to encourage engagement, but it is difficult to see this as a 'politicisation' of emotion.

The key narrative moments of set-up and denouement at the UN at the beginning and end of the film are both immediately preceded by more private and introspective personal reflections, with a spiritual/religious inflection – this is what makes it a truly 'personal' journey. The film starts with this, as DiCaprio recalls his 'first visual memories': of staring at a reproduction of Hieronymus Bosch's triptych *The Garden of Earthly Delights* (details of which are shown in close-up), which, he tells us, hung 'above my crib'. To explain this unusual choice of nursery decoration, DiCaprio tells us about his father's background in the 1960s artistic counterculture, a sequence illustrated with brief stills of images of, and/or works by, famous people his father knew (Andy Warhol, Lou Reed, Robert Crumb), and also by three black-and-white family photos, showing DiCaprio Sr. holding baby Leo, a picture of him playing with toddler Leo, and another shot of both parents holding infant Leo. These intimate photographs – the only autobiographical images in the film – help to establish the importance of the painting to DiCaprio's personal history. He then describes the story told by the painting's panels as a way to frame the issue of climate change:

> In the first panel you have Adam and Eve in the Garden of Eden. Birds flying off into the distance, elephants and giraffe and a lot of religious iconography. The second panel is where it starts to become more interesting. The deadly sins start to infuse their way into the painting. There's overpopulation, there's debauchery and excess. And the last panel, which is the most nightmarish one, especially from a young child's perspective, is this twisted, decayed, burnt landscape. A paradise that has been degraded and destroyed.

This marks the formal beginning of the film as the opening titles appear on screen, and DiCaprio's final words above are illustrated and reinforced with a fast-paced montage of images of nature being 'degraded and destroyed', including smoking industrial chimneys, forest fires, flooding and melting polar ice, accompanied by music and sound-bites from reports of climate change. As the music reaches a climax, the sequence finishes with Margaret Thatcher's voice saying 'It is these activities that drove man out of the Garden of Eden'. Right from the opening moments of the film, then, an interconnection is established between DiCaprio's personal memories and feelings and a Christian condemnation of human despoliation of God-given nature. It is this combination which frames the issue of climate change in emotional terms.

The pay-off of this emotional narrative comes near the end of the film, just before we see DiCaprio delivering his second address to the UN, and just after the most poignant of the interviews with experts that he conducts. The latter is a conversation with Dr Piers Sellers, a former astronaut and director of the Earth Sciences Division at NASA's Goddard Space Flight Center, which takes place in front of a video wall showing spectacular computer simulations of climate change. Sellers emphasises vulnerability, revealing that he has recently been diagnosed with terminal cancer (he died in December 2016), and describing the atmosphere as a 'tiny little onion skin around the Earth…an astonishingly fragile film'. This idea is picked up at the end of the interview, as the film cuts from an image of Earth from space on the video wall to the Bosch triptych, which then closes to show its exterior, depicting the earth enclosed in a glass sphere. DiCaprio explains:

> I still think a lot about that picture that used to hang above my crib. The story of *The Garden of Earthly Delights* actually begins on the outside of the painting, where Bosch painted a view of our Earth on the third day of creation. It's almost as if he wanted to show the fragility of our planet by depicting the Earth and its atmosphere enclosed in glass.

As this suggests, the return to the painting re-establishes the connections made at the beginning of the film, which are then further elaborated in the sequence that immediately follows, in which DiCaprio meets Pope Francis, the final encounter on his journey.

The meeting with the Holy Father is in some ways similar to others in the film, but it is also marked out as special. DiCaprio meets a number of powerful and important people – including US Secretary of State John Kerry, President Barack Obama, and UN Secretary General Ban Ki-Moon – but in this meeting he appears nervous, shy and awed. And while there are many 'behind-the-scenes' glimpses of DiCaprio – on the set of *The Revenant*, or chatting with Ban Ki-Moon before making a UN speech, for example – the episode with Pope Francis is the most unguarded and personally significant. We see him in a taxi on the way to the Vatican nervously practicing what he will say, for instance, and can observe his deferential demeanour. The significance of the moment is also underscored by the fact that he presents the Pope with a book on Hieronymus Bosch, showing him *The Garden of Earthly Delights* and telling him what we already know: that 'it was hanging above my crib as a young boy'. In case we were in any doubt, DiCaprio tells us that 'Being able to spend time and sit with the Pope was a pretty profound experience'. The personal importance of the experience for DiCaprio reinforces what he understands as the importance of the Pope's 2015 encyclical on climate change, 'On Care for Our Common Home':

> This is a direct message from the Pope. It's a huge deal. One of the most important spiritual leaders on the planet. He has now called upon the world community to accept the modern science of climate change. A Pope has never done anything like this in history.

After the meeting, DiCaprio recounts what the Pope said to him, telling us that 'There was definitely an urgency in his voice'. We then see images from the Bosch painting again as DiCaprio reflects on its significance one final time:

> After everything I've seen, it's become pretty obvious that we're no longer living in that first, unspoiled depiction of Eden. We're in that second panel. What Bosch called Human Kind Before the Flood. And what haunts me the most is that last panel. The one with the charred, blackened skies. A planet that we collectively have left to ruin.

As he delivers these lines, the film cuts between images from the painting and scenes of DiCaprio entering the UN building for the Paris Climate Accord signing. With the close connections between personal memory and emotion, religious or spiritual meaning, and climate change re-established, we are now ready for the narrative resolution of DiCaprio's closing UN speech.

The Vatican sequence starts with the Pope saying: 'This is our home. It is going to ruin. And that harms everyone'. DiCaprio then reiterates this, quoting from the encyclical: 'Our common home has fallen into serious disrepair'. There is no indication that he is quoting these words, which he speaks as if they are his own.[20] What does become obvious, however, is that the overall message from DiCaprio's journey is essentially a reiteration and reinforcement of the message that Pope Francis has delivered. DiCaprio has already recounted what the Pope told him in their meeting:

> He said that as far as the Paris conference is concerned, he, he felt it was a step in the right direction, but certainly not enough. He feels we all need to keep speaking out about this issue as loud as we can, and that we must immediately take action. But more than anything, he said to pray for the human race.

In his final UN speech, DiCaprio then echoes these points, saying that while an 'historic agreement' has been reached at the Paris summit, 'unfortunately the evidence shows us that it will not be enough', and it will 'mean absolutely nothing' if leaders do not have the 'sense of urgency' to push beyond it. While he does not ask his listeners to pray for humanity, he does call for a 'new collective consciousness', a 'new collective evolution of the human race', and tells the assembled world leaders that as the 'last, best hope of Earth' (quoting Abraham Lincoln), 'We ask you to protect it, or we, and all living things we cherish, are history'. This is the final, dramatic line of the film, which ends with a cut to black on a musical crescendo as DiCaprio delivers these words.

It is a powerful climactic moment, grounded in the film's building sense of personal, emotional, and spiritual meaning. Yet rather than seeing it in political terms, it makes more sense to understand the emotional pull of the film as working therapeutically, offering a semi-secularised sense of religious transcendence and meaning. As for Gore in *An Inconvenient Truth*, childhood memories linked with nature are invoked, but here they are set within a religious framework and without the positive connotations of remembered harmony with nature. Instead, DiCaprio dwells on his early memories of 'nightmarish' scenes from the painting, and his fear that the 'deadly sins' of humanity are producing degradation and destruction. Climate change is viewed through the frame of religious guilt rather than political engagement, but for a broader audience and a secular society this is rendered in therapeutic terms relating to personal doubts and feelings. The journey that DiCaprio undertakes does not make the fear go away – in his final UN speech he says 'All that I have seen and learned on my journey has absolutely terrified me' – but it dissolves his doubts (which were about himself, and his own capacity to be a messenger on climate change), and his pessimism that action could be taken. Perhaps a sympathetic viewer would find that it also resolved her doubts and gave her hope too, but it would be a stretch to describe this as a politicisation of emotion. Politically, what happens is that a wealthy Hollywood actor, inspired by the head of the Catholic Church, calls on world leaders to implement an agreement they have just signed, and tells them that they are the 'the last best hope of Earth'. It seems more like an example of Littler's (2008: 248) description of celebrity campaigning as 'plugging the gap' of legitimacy.

Conclusion: celebrity emotional correctness

Brockington's talk of the 'contradictions of capitalism', 'alienation' and 'false consciousness' (2008: 562) gives his argument about the appeals of environmental celebrity activists a faintly Marxist flavour, but rather than analysing the alienation of humanity from itself under social relations of exploitation, his approach is more like a romantic critique of capitalism, whereby 'urban living', or as Alison Anderson (2011: 6) puts it, the 'isolation and rootlessness within contemporary urban industrial society', prevents us from communing with 'nature' or 'the wild'. Contrasting the present with a bygone era when 'environmental protest demanded radical change of modern capitalism', Brockington argues that increased celebrity advocacy of environmental causes is evidence of the contemporary mainstreaming of certain varieties of environmentalism which have 'strong attachments to the dominant social and economic forces in society' (2008: 551, 553). He considers various ways of conceptualising celebrity culture – as a top-down imposition of 'false consciousness'; or conversely in a more bottom-up way, as a 'tonic demanded by estranged masses'; or in a more Foucaultian way, as a set of privileged 'discourses and semiotic regimes' which work to 'control populations' (2008: 562). There is an echo of Marx's critique of religion here,[21] and indeed Brockington's discussion is based on the work of Chris Rojek, who draws an extended parallel between celebrity culture and religion (see Rojek 2001, chapter 2).

Although it is not developed further, this perhaps hints at the strong normative dimension of celebrity climate activism. This is true in a double sense: in terms of the particular cause, and also in terms of the mode of celebrity campaigning. First, Meyer and Gamson's description of celebrities as trend-followers who prefer non-controversial, consensual issues, or Boykoff and Goodman's idea that celebrities engage with climate change to demonstrate that they are on a righteous path to 'redemption', suggests that there is a kind of 'political correctness', for want of a better term, to the choice of issue. The very prominence of celebrity environmental campaigning indicates that it is a way of demonstrating conformity to the sort of socially acceptable, environmentally correct behaviour described by Berglez and Olausson (2014: 65) that we discussed in Chapter 3. Second, we can now add to this the 'emotional correctness' (Hume 1998) of celebrity culture. As Heather Nunn and Anita Biressi (2010: 49–50) argue, celebrities play a significant role as 'emotional labourers', engaging in what they call (following Arlie Russell Hochschild 2003) 'emotion work': 'work requiring one to perform the "right" feeling and ultimately even "feel" the right feeling according to the rules of the setting'.[22] Encouraging the 'right' feelings is more like modelling an etiquette of emotion than encouraging 'politicisation'. What the effects and outcomes of such emotional modelling might be is the focus of the next chapter.

Notes

1   The 2016 US presidential election provided a striking illustration: Hillary Clinton gained a huge number of prominent celebrity endorsements (Molloy and Hod 2016), including a star-studded 'Super-PAC' called 'Save the Day' established by Hollywood writer/director Joss Whedon, who reportedly also personally donated $1m to her election campaign (Jarvey 2016); while Donald Trump, described by *Time* magazine as the 'first reality TV president' (Nesbit 2016), appeared to benefit from his fame as the host and co-producer of the television show *The Apprentice* (Heritage 2016).

2   See for example the Environmental Policy Alliance's 'Big Green Radicals' project: www.biggreenradicals.com.

3   Thompson's interview is available at http://www.bbc.co.uk/news/entertainment-arts-34135679. For the BBC Trust report, *Making Sense of Statistics*, see http://downloads.bbc.co.uk/bbctrust/assets/files/pdf/our_work/stats_impartiality/report.pdf.

4   See https://www.looktothestars.org/about.

5   See http://www.vanityfair.com/magazine/2006/05/contents-200605.

6   See https://web.archive.org/web/20161003175440/http://globalcoolfoundation.org/.

7   The other climate celebrities discussed in this chapter strike a similarly self-deprecating note through humour. In *An Inconvenient Truth*, Gore introduces himself with a jokey reference to his defeat in the 2000 presidential election ('I'm Al Gore. I used to be the next president of the United States'); while in her short film for Greenpeace, discussed below, Emma Thompson draws attention to the gap between her individual actions and

the scale of the problem she is attempting to address, for example joking that she has 'ten minutes to save the Arctic'.

8 *Why They Care*, http://yearsoflivingdangerously.com/video/why-they-care/.

9 A Q&A with Leila Conners Petersen and Nadia Conners, *Spirituality and Practice* (no date), www.spiritualityandpractice.com/films/features/view/17250.

10 Artist Tom Van Sant, whose GeoSphere image of Earth features in the film, is 'a friend of mine'; astronomer Carl Sagan is 'My friend the late Carl Sagan'; Carl Page's photograph of Mount Kilimanjaro is taken by 'a friend of mine'; 'Another friend of mine Lonnie Thompson studies glaciers'; glaciologist John Mercer is 'a friend of mine'; Harvard oceanographer James Harvey is 'a friend of mine'; and 'I had a college professor named Roger Revelle who was the first person to have the idea to measure the amount of carbon dioxide in the earth's atmosphere'.

11 *Climate Change and Arctic Oil with Emma Thompson*, 10 September 2014, https://www.youtube.com/watch?v=fFNqKA8VBGU.

12 See for example her interviews with the BBC's *Andrew Marr Show*, 21 September 2014, and with Sky News, 29 November 2015. In both, she exhorts viewers: 'inform yourself'.

13 Gaia Wise suggests that 'In 60, 70 years, when it warms 8 degrees, this is going to be gone'; and Emma Thompson notes that 'The warming process is accelerated here on top of the world. Whereas in other parts of the world we've warmed by .4, .6, .8 of a degree, here it's one degree. One full degree. Disaster.'

14 Why is Emma Thompson in the Arctic?, *The Real News Network*, 23 August 2016, http://therealnews.com/t2/index.php?option=com_content&task=view&id=31&Itemid=74&jumival=17052.

15 Journey's End: No More Self-Indulgent Celebrity Travel Shows, Please, *Guardian*, 15 February 2010, www.theguardian.com/tv-and-radio/tvandradioblog/2010/feb/15/sue-johnston-shangri-la-travelogue.

16 See *Frack Free Bake Off*: https://www.youtube.com/watch?v=bpzzFwupHHg.

17 Gabriel Huddleston (2017) has analysed how the show works through 'pedagogical empathy and kindness as affect'.

18 See http://yearsoflivingdangerously.com/watch/behind-the-scenes/, and http://yearsoflivingdangerously.com/watch/exclusives/.

19 Leonardo DiCaprio (UN Messenger of Peace) at the opening of Climate Summit 2014, https://www.youtube.com/watch?v=vTyLSr_VCcg.

20 The passage from the Pope's encyclical (para. 61) that DiCaprio quotes is: 'our common home is falling into serious disrepair. Hope would have us recognize that there is always a way out, that we can always redirect our steps, that we can always do something to solve our problems. Still, we can see signs that things are now reaching a breaking point' (http://w2.vatican.va/content/francesco/en/encyclicals/documents/papa-francesco_20150524_enciclica-laudato-si.html).

21 As Marx (1844) famously put it: 'Religion is the sigh of the oppressed creature, the heart of a heartless world, and the soul of soulless conditions. It is the opium of the people'.

22 Hochschild actually draws a distinction between emotion work and emotional *labour*, reserving the latter term to mean 'the management of feeling to create a publicly observable facial and bodily display'. This emotional labour is 'sold for a wage and therefore has *exchange value*', whereas emotion work refers to 'these same acts done in a private context where they have *use value*' (Hochschild 2003: 7n, original emphasis). The blurred line between authenticity and artifice in the activities of celebrity campaigners may render this point moot, but it seems to make more sense to understand their activities as emotional *labour*.

References

Anderson, Alison (2011) Sources, media, and modes of climate change communication: The role of celebrities, *WIREs Climate Change*, 2 (4): 535–46.

Anderson, Alison (2013) 'Together we can save the arctic': Celebrity advocacy and the Rio Earth Summit 2012, *Celebrity Studies*, 4 (3): 339–52.

Beattie, Geoffrey, Laura Sale and Laura McGuire (2011) An inconvenient truth? Can a film really affect psychological mood and our explicit attitudes towards climate change?, *Semiotica*, 187 (October): 105–25.

Berglez, Peter and Ulrika Olausson (2014) The post-political condition of climate change: An ideology approach, *Capitalism Nature Socialism*, 25 (1): 54–71.

Blühdorn, Ingolfur (2007) Sustaining the unsustainable: Symbolic politics and the politics of simulation, *Environmental Politics*, 16 (2): 251–75.

Boltanski, Luc (1999) *Distant Suffering: Morality, Media and Politics*. Cambridge: Cambridge University Press.

Boykoff, Maxwell T. and Michael K. Goodman (2009) Conspicuous redemption? Reflections on the promises and perils of the 'celebritization' of climate change, *Geoforum*, 40 (3): 395–406.

Boykoff, Maxwell T., Michael K. Goodman and Jo Littler (2010) '*Charismatic Megafauna': The Growing Power of Celebrities and Pop Culture in Climate Change Campaigns*, Environment, Politics and Development Working Paper Series (Paper 28), Department of Geography, King's College London, www.kcl.ac.uk/schools/sspp/geography/research/epd/working.html.

Brockington, Dan (2008) Powerful environmentalisms: Conservation, celebrity and capitalism, *Media, Culture & Society*, 30 (4): 551–68.

Brockington, Dan (2009) *Celebrity and the Environment*. London: Zed Books.

Brockington, Dan (2014) *Celebrity Advocacy and International Development*. Abingdon: Routledge.

Brockington, Dan and Spensor Henson (2014) Signifying the public: Celebrity advocacy and post-democratic politics, *International Journal of Cultural Studies*, DOI:10.1177/1367877914528532 [accessed electronically].

Chouliaraki, Lilie (2012) The theatricality of humanitarianism: A critique of celebrity advocacy, *Communication and Critical/Cultural Studies*, 9 (1): 1–21. DOI:10.1080/1479142 0.2011.637055 [accessed electronically].

Chouliaraki, Lilie (2013) *The Ironic Spectator: Solidarity in the Age of Post-Humanitarianism*. Cambridge: Polity.

Christiansen, Lene Bull and Birgitta Frello (2016) Celebrity witnessing: Shifting the emotional address in narratives of development aid, *European Journal of Cultural Studies*, 19 (2): 134–49.

Cronin, Melissa (2015) 'Hypocrite!' Leonardo DiCaprio took 6 private jet flights in 6 weeks Sony emails reveal – despite climate change advocacy work, *Radar*, 17 April, http://radaronline.com/exclusives/2015/04/leonardo-dicaprio-climate-change-hypocrite-sony-emails-wikileaks/.

Doyle, Julie, Nathan Farrell and Michael K. Goodman (forthcoming) Celebrities and Climate Change: History, Politics and the Promise of Emotional Witness, in Matthew C. Nisbet (ed.) *The Oxford Encyclopedia of Climate Change Communication*. Oxford: Oxford University Press, http://climatescience.oxfordre.com/page/climate-change-communication/.

Drake, Philip and Michael Higgins (2006) 'I'm a Celebrity, Get Me into Politics': The Political Celebrity and the Celebrity Politician, in Su Holmes and Sean Redmond (eds.) *Framing Celebrity: New Directions in Celebrity Culture*. London: Routledge.

Duncan, Zoey (2015) That awkward moment when you have to explain a chinook to Leo DiCaprio, *Calgary Herald*, 10 December, http://calgaryherald.com/storyline/that-awkward-moment-when-you-have-to-explain-a-chinook-to-leo-dicaprio.

Ervine, Kate (2013) Al Gore as Carbon Warrior: The Politics of Inaction, in Gavin Fridell and Martijn Konings (eds.) *Age of Icons: Exploring Philanthrocapitalism in the Contemporary World*. Toronto: University of Toronto Press.

Furness, Hannah (2016) BBC let Emma Thompson get away with 'inaccurate' climate change claims, watchdog finds, *Telegraph*, 10 August, www.telegraph.co.uk/news/2016/08/10/bbc-let-emma-thompson-get-away-with-inaccurate-climate-change-cl/.

Goodman, Michael K. (2013) *Celebritus Politicus*, Neo-liberal Sustainabilities, and the Terrains of Care, in Gavin Fridell and Martijn Konings (eds.) *Age of Icons: Exploring Philanthrocapitalism in the Contemporary World*. Toronto: University of Toronto Press.

Goodman, Michael K. (forthcoming) Environmental Celebrity, in Noel Castree, Mike Hulme and Jim Proctor (eds.) *The Routledge Companion to Environmental Studies*. London: Routledge.

Goodman, Michael K., Jo Littler, Dan Brockington and Maxwell Boykoff (2016) Spectacular environmentalisms: Media, knowledge and the framing of ecological politics, *Environmental Communication*, 10 (6): 677–88.

Gore, Al (2006) *An Inconvenient Truth*. London: Bloomsbury.

Hammond, Philip (2009) Celebrity Culture and the Rise of Narcissistic Interventionism, in Robert Clarke (ed.) *Celebrity Colonialism: Fame, Power and Representation in Colonial and Postcolonial Cultures*. Newcastle: Cambridge Scholars Publishing.

Heritage, Stuart (2016) You're hired: How *The Apprentice* led to President Trump, *Guardian*, 10 November, www.theguardian.com/commentisfree/2016/nov/10/trump-the-apprentice-president-elect-reality-tv.

Higgins, Charlotte (2007) No more private jets for me, DiCaprio tells Cannes, *Guardian*, 21 May, www.theguardian.com/film/2007/may/21/cannes2007.cannesfilmfestival.

Hochschild, Arlie Russell (2003) *The Managed Heart* (Twentieth Anniversary Edition). Berkeley, CA: University of California Press.

Huddleston, Gabriel (2017) Pedagogical empathy and kindness as affect in the Great British Bake Off, *In Media Res*, 10 February, http://mediacommons.futureofthebook.org/imr/2017/02/10/pedagogical-empathy-and-kindness-affect-great-british-bake.

Hume, Mick (1998) *Televictims: Emotional Correctness in the Media AD (After Diana)*. London: Informinc.

Isin, Engin F. (2004) The neurotic citizen, *Citizenship Studies*, 8 (3): 217–35.

Jarvey, Natalie (2016) Avengers assemble! Director Joss Whedon launches Hillary Clinton super PAC with star-studded video, *The Hollywood Reporter*, 21 September, www.hollywoodreporter.com/news/joss-whedon-launches-super-pac-931046.

Kapoor, Ilan (2013) *Celebrity Humanitarianism*. Abingdon: Routledge.

Klein, Naomi (2014) *This Changes Everything: Capitalism vs. the Climate*. London: Penguin.

Littler, Jo (2008) 'I feel your pain': Cosmopolitan charity and the public fashioning of the celebrity soul, *Social Semiotics*, 18 (2): 237–51.

Marshall, P. David (1997) *Celebrity and Power: Fame in Contemporary Culture*. Minneapolis, MN: University of Minnesota Press.

Marx, Karl (1844) A Contribution to the Critique of Hegel's Philosophy of Right, www.marxists.org/archive/marx/works/1843/critique-hpr/intro.htm.

Meyer, David S. and Joshua Gamson (1995) The challenge of cultural elites: Celebrities and social movements, *Sociological Inquiry*, 65 (2): 181–206.

Mohr, Ian and Emily Smith (2016) Hypocrite! Leo takes private jet to collect environmental award, *Page Six*, 20 May, http://pagesix.com/2016/05/20/hypocrite-leo-takes-private-jet-to-collect-green-award/.

Molloy, Tim and Itay Hod (2016) 167 Hollywood Stars for Hillary Clinton, *The Wrap*, 28 October, www.thewrap.com/hollywood-stars-for-hillary-clinton-list/.

Murray, Robin and Joseph Heumann (2007) Al Gore's *An Inconvenient Truth* and its skeptics: A case of environmental nostalgia, *Jump Cut*, 49, www.ejumpcut.org/archive/jc49.2007/inconvenTruth/.

Nesbit, Jeff (2016) Donald Trump is the first true reality TV president, *Time*, 9 December, http://time.com/4596770/donald-trump-reality-tv/.

Norgaard, Kari Marie (2011) *Living in Denial: Climate Change, Emotions and Everyday Life*. Cambridge, MA: MIT Press.

Nunn, Heather and Anita Biressi (2010) 'A trust betrayed': Celebrity and the work of emotion, *Celebrity Studies*, 1 (1): 49–64.

Powers, John (2014) Naomi Klein on *This Changes Everything*, her new book about climate change, *Vogue*, 26 August, www.vogue.com/1009011/naomi-klein-this-changes-everything-climate-change/.

Robson, Fletcher (2016) DiCaprio's Oscar speech cringe-worthy for some advocates of climate-change action, *CBC News*, 29 February, www.cbc.ca/news/canada/calgary/leonardo-dicaprio-climate-change-oscar-speech-alberta-reaction-1.3469010.

Rojek, Chris (2001) *Celebrity*. London: Reaktion Books.

Rojek, Chris (2014) 'Big Citizen' celanthropy and its discontents, *International Journal of Cultural Studies*, 17 (2): 127–41.

Street, John (2004) Celebrity politicians: Popular culture and political representation, *British Journal of Politics and International Relations*, 6 (4): 435–52.

Thrall, A. Trevor, Jaime Lollio-Fakhreddine, Jon Berent, Lana Donnelly, Wes Herrin, Zachary Paquette, Rebecca Wenglinski and Amy Wyatt (2008) Star power: Celebrity advocacy and the evolution of the public sphere, *The International Journal of Press/Politics*, 13 (4): 362–85.

Urry, John (2002) *The Tourist Gaze* (Second Edition). London: Sage.

5

CELEBRITY SOLUTIONS AND THE RADICAL ALTERNATIVE

In our discussion of celebrity campaigning in Chapter 4, we left unexamined the assumption that the primary goal and purpose of such activism is to raise the visibility of a particular cause or issue and to mobilise popular audiences who might not otherwise engage with it. Recent research has suggested that this assumption is mistaken, however, and that as Dan Brockington (2014: 9) puts it, 'Celebrity advocacy is by and for elites'. If this is correct, then it puts the debates about the celebritisation of climate change in a new light. This chapter first outlines the evidence on audience engagement with celebrity campaigning, and examines its implications for how to understand what we identified earlier as the therapeutic emotional appeal of climate celebrities. It then goes on to look at the different political and policy solutions that this appeal supports, comparing what have been criticised as the mainstream, post-political prescriptions of figures such as Al Gore, with the ostensibly much more radical alternative proposed by Naomi Klein.

Alienating the public

Critics have offered different assessments of the effectiveness and impact of celebrity campaigning, raising concerns that audience attention may focus primarily on entertainment rather than politics (Thrall *et al.* 2008: 378), on the famous person rather than the cause (Meyer and Gamson 1995: 187), or on the celebrity's feelings rather than the issue (Chouliaraki 2012: 16). They have also differed regarding whether the audience is thought of as naïve or knowing: David Meyer and Joshua Gamson (1995: 201) characterise audiences as 'often skeptical rather than gullible', for example, and Jo Littler (2008: 239) acknowledges that celebrity displays of caring are 'routinely and publicly mocked'. For Ilan Kapoor (2013: 42–4), audience scepticism coexists with enjoyment and does not break what he sees as popular complicity in the 'ideological fantasy' offered by celebrity campaigners. Even if we view it ironically, we still go along with the fantasy, allowing celebrities to 'become stand-ins for our feelings of beneficence or compassion': like the act of turning a prayer wheel or hiring professional mourners for a funeral, 'not only do others believe for us, but through this transference we no longer need to believe' (2013: 43). Notwithstanding these different views, however, there are some common, indeed common-sensical, baseline assumptions: that attaching a famous person to a cause or campaign helps to get it noticed, attracting media coverage and public attention; that this strategy works because the public are interested in celebrity; and that, in the best case scenario, this interest can be used to pull in larger audiences who would not normally engage with an issue.

Yet there are strong reasons to question these assumptions. As Tim Markham (2015: 467–8) puts it: 'whatever we might think politically or morally about celebrity advocacy, recent evidence...suggests in fairly strong terms that it does not work....it does little to raise awareness in any way that leads to public engagement in and tangible support for an issue'. One important challenge to the common-sense understanding of celebrity campaigning is Trevor Thrall *et al.*'s (2008: 364) finding that 'although most celebrities participate in various forms of advocacy, rarely do even the most famous celebrities get sustained attention from mass media news organizations for advocacy-related activity'. While the majority of the celebrities in their sample of 247 were active as advocates for a cause, only four of them generated more than a hundred news stories over a one-year period, and most of them (54.5%) received zero press coverage for their efforts (2008: 369). Thrall *et al.* also tested the possibility that celebrity campaigning might be more successful in entertainment-focused media, by including *People* magazine in their study and looking specifically at stories about celebrity environmental activism. Yet here the results were even more stark: 97% received zero coverage (2008: 371). Thrall *et al.* hypothesise that perhaps celebrity advocacy works differently, via online social media targeted to specific audiences, rather than as part of the broader public sphere (2008: 381).

It is questionable, however, just how interested people are in celebrity in relation to climate change. Max Boykoff and Mike Goodman (2009: 399–400) refer to a Nielsen/Oxford University survey which appeared to show that 'specific celebrities have garnered particular discursive sway' on the issue. Yet the survey offered a selection of 22 famous people from which respondents could choose their three most influential 'celebrity champions' on climate change. Top of the list was Al Gore, followed by former UN Secretary General Kofi Annan, with Bill Clinton and Oprah Winfrey tied for third place. Only Oprah would normally be thought of as primarily a 'celebrity', the others having come to prominence through more conventional political/diplomatic careers. The list of choices included five people in the 'politicians' category, eight actors, four sports personalities, three musicians, and two classified as 'other' (Richard Branson and Robert F. Kennedy Jr.). The survey basically found that people thought politicians were the most influential people to champion efforts to combat climate change, despite the array of other choices on offer.[1] Even more strikingly, Alison Anderson (2011: 8) notes that surveys of children and young people carried out for the UK government's Department for Environment, Food and Rural Affairs in 2006 and 2008 found that most could not identify anyone when asked to name 'celebrities who care about climate change'. In the 2008 survey, 6% identified singer Bob Geldof, but 50% answered 'don't know' and a further 20% said 'nobody', with the remainder suggesting either the prime minister, 'politicians' or 'the government'.[2] Matthew Hibberd and An Nguyen's (2013: 37) study of the effectiveness of climate change communication for young adults, including attempts to 'make climate change more relevant and appealing' by involving celebrities, found that a few participants were 'apparently positive' about such initiatives and saw them as potentially influential (on other people's opinions rather than necessarily their own). In general, though, the participants thought that 'celebritisation' was 'potentially counterproductive because celebrities are perceived to have no legitimacy in climate campaigning (for example, they are among the highest carbon emitters and they do not have the necessary scientific knowledge)' (Hibberd and Nguyen 2013: 28).

An important study by Nick Couldry and Tim Markham (2007) further challenges the assumption that audiences are interested in celebrity and that this could be used as an indirect route to get them interested in public issues, including the environment. Their study of UK media audiences found that celebrity stories were a minority interest, and that there was 'a wide range of attitudes to celebrity culture, with considerable ambivalence and sometimes hostility being found alongside attraction and engagement' (2007: 409). The minority (one in seven) who did prioritise keeping up with celebrity and popular culture were typically younger, female audience members who were the least engaged in politics, least likely to vote, least likely to get involved in local organisations or volunteer work, and least likely to participate in 'action or discussion about public-type issues' (2007: 403, 413). As distinct from the audience cluster who were simply not interested in politics and current affairs, this group of respondents made a positive choice to focus on celebrity-related stories, suggesting that 'people's engagement with celebrity culture is part of a turning *away* from concern with issues that require public resolution' (Couldry and Markham 2007: 418, original emphasis). Indeed, those attentive to celebrity were not simply uninterested in politics, they were more likely to be cynical about it, most feeling that they had no say, that politics was irrelevant to their lives, and that it made no difference who was in power (2007: 415). These findings lead Couldry and Markham not only to challenge the claim that celebrity culture 'contributes positively to the possibilities of democratic renewal', but also to wonder whether 'the "politics" of celebrity culture…[is] an academic illusion' (2007: 405, 411).

Couldry and Markham's conclusions have been confirmed and extended by Brockington's audience research on celebrities and international development (Brockington 2014, Brockington and Henson 2014). He too found evidence which made him question both the assumed popularity of celebrity culture and its supposed efficacy in connecting people with the charities and causes for which celebrities campaign. Most of Brockington's respondents engaged only briefly and superficially with celebrity news: most were either ambivalent or hostile to celebrity campaigning for development causes, and those who were interested in *celebrities* tended not to be interested in celebrity *advocacy* (Brockington and Henson 2014: 7–11). Like the respondents in Hibberd and Nguyen's (2013) study who thought that celebrities might be influencing other people, Brockington too discovered a strong third-person effect supporting 'a false, but popular belief, that celebrity is popular' (Brockington and Henson 2014: 9).

Brockington's most important point, though, is that 'celebrity advocacy, conceived as a means of involving the public, is in fact a means of its alienation' (Brockington and Henson 2014: 4). Drawing on Colin Crouch's (2004) work on post-democracy, Brockington argues that celebrity advocacy is 'bound up in, if not actively promoting, a post-democratic form of politics, with minimal actual involvement of citizens', since it is 'a means by which citizens *disengage* from agonistic politics, at the same time as political elites perceive them to be engaged in politics' (Brockington and Henson 2014: 15, original emphasis). This view so cuts against the grain of our common-sense understanding that it may need a little further explanation. First, Brockington finds, in line with other studies, that although celebrity campaigning 'occupies a significant proportion of the public domain', it does so 'without always engaging particularly well with much of the public' – indeed, the 'very act of public engagement in the form of celebrity can alienate' (Brockington 2014: 8–9).

Second, he argues that the most 'fervent believers in celebrity power' are political elites, with whom celebrity advocacy therefore works because they take celebrity involvement in a cause as an indication of wider public interest (Brockington and Henson 2014: 15).

> In other words, even though the public are not engaged by a campaign, or even by the celebrities involved in a campaign, politicians will think that they are because of those celebrities. Celebrities *signify* the public....There is, then, a rather rich irony in celebrity advocacy. It is not just that its actual popular reach is different from popular belief about that reach. It is that its very influence hinges upon that difference. If it were not misrecognized, it could not exist.
>
> (Brockington and Henson 2014: 15, original emphasis)

The presence of celebrities 'simulates public involvement' rather than actually encouraging it (Brockington 2014: 162).

Third, Brockington (2014: 8–9) argues that in a post-democratic context, 'characterised by a loss of democratic verve and the corresponding rise of government by elite', this ersatz form of public engagement entails a 'reshaping of politics according to the imagination of the elites who dominate it'. The point finds some corroboration in Chouliaraki's (2012: 2) description of celebrity humanitarianism as supporting a 'theatrical conception of politics'. In her discussion of Angelina Jolie's activities 'both in the political sphere, where she engages in the inner circles of elite lobbying, and in the financial sphere, where she is in charge of her own foundation', Chouliaraki describes a move away from meaningful public participation, mass mobilisation and grassroots action in a context of 'receding collectivities of solidarity' (2012: 13–14). It is in this 'theatrical arrangement of separation between the acting celebrities and their watching publics', that celebrities perform their 'exemplary dispositions of emotion' (2012: 3).

Despite all his crushing condemnations of celebrity campaigning as an 'integral part' of 'iniquitous' political and economic systems that produce 'profound inequality' and 'favour the interests of the wealthy', Brockington nevertheless wonders if it 'may be redeemable' and if it could 'become a force for egalitarian politics and globally fairer economic structures' (2014: 162). There seems to be little reason to think so, and indeed Brockington acknowledges that the chances are 'slim'. But the choice, as he presents it, is between an idealistic (and, he implies, unrealistic) aspiration to 'build a more vigorous civic culture and active agonistic politics', or a more 'pragmatic' decision to work within an 'elite-oriented approach' (Brockington and Henson 2014: 16; and see also Brockington 2016: 216).

Brockington poses this choice in the context of humanitarianism and development causes, but he and others have extended the argument to celebrity climate campaigning. Goodman (forthcoming: 3) notes that 'perhaps the key intervention that environmental celebrities make is when they talk to other elites', though he acknowledges that 'celebrities – and the economic and cultural systems that support them – are fundamentally un-democratic'. Likewise, in an article co-authored with Brockington, Goodman et al. (2016: 684) concede that publics 'may not be listening', but suggest that celebrity advocacy can still be effective because 'elites do notice what celebrity spokespeople say'. Similarly, Doyle et al. (forthcoming: 23–4) suggest that the key audience for the emotive 'after data' appeal of climate celebrities that we discussed in Chapter 4 'might actually not be the general public but rather other elites', and describe DiCaprio's *Before the Flood* as an example of 'elite-to-elite emotionally-tinged communication'.

This is the extraordinary *reductio ad absurdum* of the critique of celebrity campaigning: any failing, up to and including its anti-democratic and elitist character, still does not dislodge hopes that it may have an upside – even if that upside has to be located in the positive potential of the very same undemocratic elites that the critique has denounced. The pragmatic turn to 'elite-oriented' approaches suggests a deep-seated scepticism about democratic engagement, thereby adopting the same post-political logic as celebrity campaigning itself.

As Chris Rojek (2014: 134) argues in his critique of 'Big Citizen' celanthropy, today's 'technocratic' celebrity activism is inherently anti-political, since it is premised on the defects, limitations and incompetence of elected governments, and works instead through 'stateless solutions' operating outside of formal politics. It is a mode of campaigning that 'turns citizens into spectators', and subjects them to subtle forms of control and 'moral regulation' (2014: 134, 136). Interestingly, Rojek sees this 'power regime' as operating through 'various fronts of permanent external menace – terrorism, economic melt-down, the "threat" of immigration' (2014: 136) – to which we might well add environmental catastrophe. The point here, though, is that he understands the appeal to a popular audience as a 'placebo effect' (2014: 134) – producing a gratifying sense of having done something good, while actually not fundamentally changing anything. This, Rojek argues, has 'therapeutic value' for popular audiences of celebrity advocacy, providing 'ordinary people with a therapeutic sense of effective activism in the midst of routine experience of powerlessness, helplessness and emasculated democracy' (2014: 127, 136). Elsewhere, Rojek (2012: 178) has also argued that the appeal of celebrity culture is that it 'affords access to the deep human need for transcendence and meaning'. Celebrities offer a 'desacralized highway to transcendence', and 'provide millions of people with a sense of meaning', he argues (2012: 1). Rojek may or may not be right about how 'ordinary people' view celebrity, but what would his argument imply if we take on board the idea that the key audience for campaigning celebrities is the elite?

Heather Nunn and Anita Biressi observe that the celebrity performance of emotion is part of a larger phenomenon: it 'intersects with the broader cultural imperative to emotion work as the foundation of a productive and successful life in the public *political* realm' (2010: 59, original emphasis). Such is the expectation that public figures will be 'willing to engage in the business of emotional intimacy' that the convincing demonstration of 'authentic feeling' has become a requirement for gaining public trust (2010: 56–7, 49). This leads Nunn and Biressi to endorse Barry Richards's (2007) idea that in a therapeutic public sphere, it is the emotionally literate who will be best able to engage the electorate, or as they put it: 'it will be the public figures who best understand the emotional dimensions of the public realm and its communicative rules who will most effectively connect with their publics' (Nunn and Biressi 2010: 62). For Richards (2004: 340), the 'emotionalisation' of politics results from its having become intertwined with popular culture. His reasoning is that since 'popular culture is substantially about feeling, about the expression and management of emotion', so the 'incursion into political experience of the values of popular culture means that we now seek certain kinds of emotionalized experience from politics'. This provides both an explanation for, and in his mind a possible solution to, the 'democratic deficit': popular disengagement from conventional politics is understood as resulting, at least in part, from an 'emotional deficit in political communications – that is, the failure of these communications to satisfy the contemporary taste for certain kinds of affective experience'. In response, politics needs to acquire 'something of the emotionally compelling narratives offered by, for example, television soap operas' (2004: 340, and see further Richards 2007).

Although Richards suggests that the emotionalisation of politics could be described in terms of a *'consumer* culture in which there is a cultivated prominence of feeling' (2004: 340, original emphasis), he generally understands it as working in a bottom-up way, as a 'spontaneous movement in culture' (2007: 42). Yet Brockington's work implies that the demand for therapeutic transcendence is a top-down phenomenon. Borrowing or imitating the intimate, confessional style of celebrity culture has become a way for political leaders to offset the crisis of legitimacy and popular suspicion of conventional formal authority that characterises contemporary public life (Furedi 2010).

As we saw in Chapter 4, DiCaprio's 'elite-to-elite emotionally-tinged communication' (Doyle *et al*. forthcoming: 23–4) does indeed offer a kind of therapeutic transcendence in the way that Rojek describes, but perhaps it is at least as much for elites as for any wider audiences. Perhaps after all it is the people in the room, those he is directly addressing – whether political leaders at a UN summit, fellow actors at the Oscars ceremony, business leaders at the Davos World Economic Forum, or bankers and minor royalty at the annual Saint-Tropez fundraising galas of the Leonardo DiCaprio Foundation – who are the key audience, rather than the spectators watching via a screen.[3] Perhaps, rather than seeing emotive celebrity appeals as endowing the lives of powerless 'ordinary people' with meaning, a more pressing and proximate deficit of meaning is that of the global elites for whom celebrities simulate public involvement.

Certainly for a celebrity activist politician like Gore, there appears to be an urgent imperative to gain a sense of 'transcendence and meaning' (Rojek 2012: 178). As he explains:

> The climate crisis also offers us the chance to experience what very few gen-
> erations in history have had the privilege of knowing: *a generational mission*;
> the exhilaration of a compelling *moral purpose*; a shared and unifying *cause*; the
> thrill of being forced by circumstances to put aside the pettiness and conflict
> that so often stifle the restless human need for transcendence; *the opportunity
> to rise*. When we do rise, it will fill our spirits and bind us together. Those who
> are now suffocating in cynicism and despair will be able to breathe freely.
> Those who are now suffering from a loss of meaning in their lives will find
> hope. When we rise, we will experience an epiphany as we discover that this
> crisis is not really about politics at all. It is a moral and spiritual challenge.
>
> (Gore 2006: 11, original emphasis)

This is from the introduction to the book version of *An Inconvenient Truth*, which, from the descriptions of the film discussed in Chapter 4, one might expect to be even more dry and data-heavy. Yet it turns out that Gore is actually in pursuit of a spiritual 'epiphany'. This is not a politicisation of emotion – Gore explicitly disavows politics here – but rather an attempt to fill a politics-shaped hole with emotional therapeutic transcendence.

Celebrity solutions

A final aspect of the critique of celebrity advocacy which needs to be re-examined is the scope of the proposed solutions that it typically involves. As Doyle *et al*. (forthcoming: 28) observe, even if celebrities were to succeed in getting people to be 'more emotive about climate change', this would count for little if 'the solutions celebrities propose include the typical "weak brew" of more and better conscious capitalism, sustainable consumption and individual responses of light-bulb changing'. As we have seen, focusing on minor lifestyle changes on the part of responsibilised individuals is a frequent target of criticism, and one that DiCaprio acknowledges in *Before the Flood*. Recalling his involvement with Earth Day 2000, he comments that:

> Back then everyone was focused on small, individual actions…it boiled down to simple solutions like changing your light bulb. It seemed like a positive thing at the time, you know, changing your light bulb. But it's pretty clear that we're way beyond that point now. Things have taken a massive turn for the worse.

This is not to say that the film does not raise issues of lifestyle and consumption: it includes sections on palm oil production and beef farming, for example, which largely focus on individual consumer choices. In general, though, the emphasis is on much larger-scale action, to be enacted via the UN, national governments and businesses: in particular, the film features Harvard Economics Professor Gregory Mankiw advocating carbon taxing, and Tesla/SpaceX CEO Elon Musk promoting solar energy.

Of course, even if such strategies for tackling climate change avoid an individualistic and consumerist focus, they stay within the logic of contemporary market society. Indeed, in view of our discussion of the intra-elite dynamics of celebrity campaigning we could understand DiCaprio's advocacy of taxation reform and government-subsidised renewables as part of an economic and political competition between different sections of the capitalist class – what Noam Chomsky has described as a division between 'high-tech, capital-intensive, internationally-oriented business, which tends to be what's called "liberal"; and lower-tech, more nationally-oriented, more labour-intensive industry, which is what's called "conservative"'.[4]

A somewhat similar picture emerges from Joel Kotkin's (2014) analysis of an historic socio-economic shift that has taken place in the last few decades in the US, involving the rise to power of two new important elite groups, which he dubs the Oligarchy and the Clerisy. The Oligarchs are the new super-rich, whose wealth and power is based in the increasingly important high-tech digital economy, mostly located around Silicon Valley and Seattle. Their political influence and patronage is largely directed toward liberal political causes, and in this their outlook and values coincide with those of the Clerisy – the intellectual elite in academia, government, the media and non-profit sectors, whose power derives from 'persuading, instructing, and regulating the rest of society' (Kotkin 2014: 8). Although there are noteworthy conservative factions, the contemporary Clerisy is overwhelmingly 'socially liberal, environmentally self-conscious, and globally-oriented' (2014: 47). Members of the Clerisy are 'increasingly uniform in their worldview', and their wider enforcement of intellectual conformity, Kotkin argues, is 'particularly notable in the area of climate change, where serious debate would seem prudent not only on the root causes and effects but also on what may present the best solutions' (2014: 8, 57). For the Oligarchs too, he suggests, environmentalism is the 'most critical part of the tech political agenda':

In their embrace of a strong opposition to fossil fuels, the Oligarchs are facing off against some of the nation's most powerful, long-established plutocratic interests and wealthiest individuals. This assault on traditional energy reflects in part the relative lack of sensitivity by many tech firms to high electricity prices.

(Kotkin 2014: 38)

A good illustration is the tech entrepreneur Carl Page (whose photo of Mount Kilimanjaro features in *An Inconvenient Truth*). Brother of Google co-founder Larry Page, Carl is founder and president of the Silicon Valley environmental NGO The Anthropocene Institute and on the board of climate think-tank ecoAmerica, and in that capacity advised Hillary Clinton's presidential campaign on how to pitch their message on climate change. Commenting on a campaign video on the topic, Page noted some 'big errors in the film that could be improved', starting with the fact that it failed to mention 'the #1 reason for why companies like Facebook and Google and Apple use so many renewables', namely that this is 'Energy that always gets cheaper'. The 'self-sacrifice austerity, saving the grandkids argument', Page noted, was 'weak'. Instead, in order to convince 'conservative businesses, and Americas [sic] people', the 'right argument is pro-prosperity!'[5]

It seems unlikely that high-tech innovations in renewable energy of the sort pursued by Musk, Page and others are solely motivated by disinterested altruism, and they are obviously not anti-capitalist or anti-consumerist. The businesses of the new Oligarchs tend to employ far fewer workers compared with traditional, 'tangible' industries, Kotkin (2014: 7, 41) observes, and often outsource production, moving it overseas. This has not only insulated them from the sorts of criticism levelled at the 'one per cent' in other sectors such as banking and finance, it has even allowed them to cultivate an anti-capitalist image. The Occupy Wall Street activists paused their protesting to mourn the passing of Apple CEO Steve Jobs in 2011, for example (Kotkin 2014: 25). Yet considered as capitalist employers, the cleaner, greener industries of the new Oligarchs exploit workers just like in any other capitalist enterprise. In *Before the Flood*, Musk not only advocates government support for 'sustainable energy', he also calls for a 'popular uprising' against the money and influence of the fossil fuel industry. Yet workers in Musk's own companies, which already receive substantial government subsidies ($4.9 billion in 2015, according to the *Los Angeles Times* (Hirsch 2015)), complain of working longer hours for lower wages, compared to the auto-industry average, incurring injuries 'often', and being required to sign a confidentiality agreement that 'threatens consequences if we exercise our right to speak out about wages and working conditions' (Moran 2017). Tesla workers have reportedly passed out from pain and fatigue while co-workers are directed to simply work around them (Wong 2017).

For some more radical critics, the failure to take an anti-capitalist stance is the core problem with celebrity climate activism and other forms of mainstream environmentalism. Kate Ervine (2013: 110–11), for example, excoriates Gore as an 'eco-icon' who uses his 'rock-star fame' to promote pseudo-solutions: the cap-and-trade system (whereby a company's maximum allowed CO_2 emissions are capped by government, but can be augmented by buying carbon credits), and the associated practice of 'carbon offsetting' (whereby businesses or individuals can 'offset' their own carbon use through funding projects such as tree-planting elsewhere, and thereby achieve 'carbon neutrality'). These strategies, institutionalised internationally since the 1997 Kyoto Protocol agreement, are ineffective, prone to gaming and fraud, and create the impression of action while actually bringing about little or no change, she argues (2013: 104–9). Gore's celebrity-style campaigning is hugely effective in one sense – Ervine (2013: 100, 110) describes him as 'adored' by 'devout followers' who accord him a 'saintly status, worshipping a prophet' – but ultimately only in negative, counter-productive ways, since it proposes a form of 'market environmentalism' which sanctions further promiscuous consumption as a 'path to green deliverance'. The overtone of religious transcendence in Ervine's description of Gore is, as suggested above, entirely appropriate. Yet even if a celebrity campaigner like Gore succeeds in attracting public attention and support, in other words, this is actually quite disastrous if the 'solutions' on offer allow the continuation of business as usual. However, there is at least one celebrity activist, Naomi Klein, whose work has been welcomed as avoiding the limitations often associated with celebrity campaigning and as offering a radical alternative (see, for example, Ervine 2015, Gunster 2017).

The radical alternative

Klein's writings and her film, *This Changes Everything*, not only reject 'market environmentalism' as a response to climate change, but take an explicitly anti-capitalist position, highlighting the effects of climate change and fossil fuel industries on people in 'sacrifice zones' of high pollution and poverty. It is an issue, she argues, that in one sense conservative climate sceptics have understood better than progressives and left-wingers: 'denialists' may be 'dead wrong about the science', she says, but they grasp the 'revolutionary meaning of climate change' (Klein 2011). Unlike many 'professional environmentalists' – who 'paint a picture of global warming Armageddon, then assure us that we can avert catastrophe by buying "green" products and creating clever markets in pollution' – anti-environmentalist think-tanks such as the Heartland Institute understand that responding effectively to climate change would entail 'deep changes...not just to our energy consumption but to the underlying logic of our economic system' (Klein 2011). As she puts it in her film, 'if climate change is taken seriously, it changes everything'.

In terms of specific policy proposals, Klein advocates breaking 'every rule in the free-market playbook':

> We will need to rebuild the public sphere, reverse privatizations, relocalize large parts of economies, scale back overconsumption, bring back long-term

> planning, heavily regulate and tax corporations, maybe even nationalize some of them, cut military spending and recognize our debts to the global South.
>
> (Klein 2011)

While such policies would require state intervention and top-down regulation, Klein emphasises the prior importance of bottom-up action and grassroots protest, which feature extensively in her documentary. One example is Germany, where 'the government intervened big time', adopting energy policies that have resulted in 'one of the most rapid energy transitions [to renewables] anywhere in the world'. As Klein acknowledges elsewhere, this may be an over-optimistic assessment,[6] but the point here is the political lesson that she draws from it: that this change 'didn't happen because the government just saw the light', but was brought about by action from below. The German people 'didn't wait for a leader, they did it on their own', sparking a 'true power shift' by taking the 'electricity grid back from private corporations' and running it themselves 'often through democratic cooperatives'. Just as Klein advocates far-reaching political and socio-economic change but sees this primarily as happening in a bottom-up fashion, so her advocacy of renewable energy does not look for solutions in large-scale, high-tech innovations but in 'community-controlled renewable energy projects, in community-supported agriculture and farmers' markets, in economic localization initiatives…and in the co-op sector' (Klein 2011).

Klein (2011) offers a 'green–left worldview', arguing that climate change can be the 'catalyst' for a 'profound social and ecological transformation' because it 'supercharges the pre-existing case for virtually every progressive demand on the books, binding them into a coherent agenda based on a clear scientific imperative'. The crisis has now reached the point where mild, moderate, mainstream solutions are inadequate, she argues: what is required is radical systemic change. This makes climate change a 'historic opportunity', a 'political game changer' (2014: 10, 154). Yet, as her localism suggests, the worldview she advocates is actually much more green than red; left-wing in rhetorical form, but ecological in political content. That is to say, rather than identifying the social relations of capitalism as the core problem to be addressed, Klein understands the issue as an 'expansionist, extractive mindset' and the 'overconsumption of natural resources' (Klein 2011). In the film, she identifies the problem as 'growth' or 'the fossil fuel economy', and presents climate change as pitching 'all of us against the logic of domination and growth at any cost'. In the book on which the film is based, she explains that 'This is a shift that challenges not only capitalism, but also the building blocks of materialism that preceded it, a mentality some call "extractivism"' (Klein 2014: 25). Not unlike Brockington's (2008: 562) discussion of alienation from nature resulting from 'urban living', noted in Chapter 4, this is a romantic critique of capitalism. Indeed, Klein (2014: 157) suggests that climate change is 'the most powerful argument against unfettered capitalism since William Blake's "dark Satanic Mills" blackened England's skies (which, incidentally, was the beginning of climate change)'.

There is no room here for the modernist ambition of Musk's solar gigafactories and futuristic plans for space exploration. That sort of thinking is rejected as 'technological hubris' (2014: 75) and she devotes a chapter of *This Changes Everything* to critiquing the idea of large-scale geo-engineering projects (2014, chapter 8). As Klein acknowledges, ecological thinking is also a 'profound challenge' to the 'extractivist left' (2014: 176–8). Whereas historically the left-wing alternative was generally about increasing material wealth and prosperity, with capitalism understood as a barrier to such progress, in Klein's argument it is modernity itself that is the problem. Hence she argues that 'while contemporary, hyper-globalized capitalism has exacerbated the climate crisis, it did not create it' (2014: 159). Rather, the problem can be traced back even earlier than the industrial revolution, to early-modern science: the seventeenth-century philosopher Francis Bacon is the 'patron saint' of extractivism (2014: 170). This is an argument which, as she notes, is 'profoundly challenging…for all of us raised on Enlightenment ideals of progress, unaccustomed to having our ambitions confined by natural boundaries' (Klein 2011). Once the problem is understood in this way – as rooted in Enlightenment ideas about scientific knowledge and human progress – the necessity of overcoming social limits is replaced with an imperative to respect natural limits.

Interestingly, this idea is linked with the personal and emotional aspects of Klein's climate writing. As Goodman (forthcoming: 1) notes, Klein has joined the ranks of celebrities who 'witness and "emote" for audiences', and has made a short film for the *Guardian*, in which her then four-year-old son travels with her to witness the impact of climate change on Australia's Great Barrier Reef while Klein discusses her 'emotional responses'.[7] Klein herself explains that she decided to bring her son into her public campaigning because of the need to 'get at aspects of climate disruption that scientific reports and political arguments just can't convey': this is 'intensely emotional terrain', yet 'climate discourse is usually pretty clinical, weighed down with statistics and policy jargon' (Klein 2016). While 'information is important', she argues, in order to be effective 'we are all going to have to stop being so impeccably calm and reasonable', and instead 'find that part of ourselves that feels this threat in our hearts, as well as our heads' (Klein 2016).

It is not unusual, as we saw in Chapter 1, for children to become a focus for fears and hopes about the environment, but in *This Changes Everything* this is intertwined particularly closely with Klein's thinking about ecology. She says that she used to look away, but now 'I try and feel it' because 'I owe it to my son' (2014: 28). In the introduction to the book, explaining the 'personal reasons' why it was difficult to write, she says that 'What gets me most are not the scary scientific studies…It's the books I read to my two-year-old', which depict a natural world that she fears he may never see (2014: 26).

Even more importantly, Klein's understanding of the environment is bound up with her difficult personal experiences of trying to have a child. First she postpones having children because of 'ecological despair', and then, after two miscarriages, she embarks on an unsuccessful course of fertility treatment, a 'technological fix' (2014: 420–21). She feels 'uneasy' at the fertility clinic because of 'many of the same things that made me wary of the geoengineers': as her personal life becomes filled with 'failed pharmaceutical and technological interventions', the ecological crisis and her personal 'fertility crisis' become interrelated (2014: 422–3). The earth is 'a mother facing a great many fertility challenges of her own', and Klein starts to think that 'protecting and valuing the earth's ingenious systems of reproducing life' requires a 'shift in worldview', from 'domination and depletion' to 'regeneration and renewal' (2014: 424). Blaming a further miscarriage on time spent reporting on the aftermath of the 2010 BP oil spill in Louisiana, Klein sees that climate change is 'depriving life-forms of their most essential survival tool: the ability to create new life'. Instead, 'the spark of life is being extinguished, snuffed out in its earliest, most fragile days: in the egg, in the embryo, in the nest, in the den' (2014: 434). She turns to a naturopathic doctor, adopting a new diet and lifestyle, moving to the countryside and experimenting with acupuncture and Chinese herbs.

In case anyone has missed them, Klein spells out the 'parallels between my new doctor's theories about infertility and some of the ideas I was encountering about the changes humanity must make if we are to avoid collapse', leading her to fresh insights into 'the limits beyond which life cannot be pushed', and a deeper appreciation of concepts of 'balance, or harmony, common to many Indigenous cultures' (2014: 438, 442–3). Ultimately, the lesson she draws is that 'what matters is that we are acknowledging that we are not in charge, that we are part of a vast living system on which we depend' (2014: 444).

This personally meaningful aspect of Klein's argument is not an add-on extra: it is central both to the book and film of *This Changes Everything*. The leading idea is that, as she says at the beginning of the film, 'the real problem is a story, one we've been telling ourselves for 400 years'. This is the story of human 'dominion over nature'; a '400-year-old fantasy' that was made to seem credible by fossil fuels. Now, however, climate change is 'telling us to stop pretending we can control nature and start acting like we are nature'. We have lost touch, she explains, with the ancient wisdom preserved in the cultures of Indigenous peoples (who feature prominently in the film). But solar power and other forms of renewable energy could herald 'the rebirth of that other story, one that all humans used to know', since renewables carry the implication that 'we can't escape or control nature. We have to work with it'. Similarly, in the book (2014: 159) she explains that 'the roots of the climate crisis date back to core civilizational myths on which post-Enlightenment Western culture is founded – myths about humanity's duty to dominate a natural world that is believed to be at once limitless and entirely controllable'. We need to reject those myths, she argues, to give up the 'illusion of total power and control', and 'accept (even embrace) being but one porous part of the world, rather than its master or machinist, as Bacon long ago promised' (2014: 175).

In this 'battle of cultural worldviews', Klein's overriding point is that 'we are not apart from nature but of it' (2014: 61). The parallels she draws between her own experience and that of the earth, particularly in terms of the need for alternative, healing narratives rather than further domination by 'techno-science', suggest that what first appears to be a radical political argument may be more therapeutic in character.

Klein's ideas closely mirror the argument made by Kate Ervine regarding Gore's mainstream environmentalism. The fundamental problem, Ervine suggests, is the place assigned to nature in 'a particular narrative of *who we are*' (2013: 95, original emphasis). Western societies separate nature as 'ontologically distinct' from humanity; as 'that which must be tamed, dominated and controlled' (2013: 96). Drawing on the work of Bruno Latour, as well as that of the 'risk society' theorists Anthony Giddens and Ulrich Beck, Ervine (2013: 96) argues that 'the emergence of environmental crises...strikes at the heart of our own self-identification as modern and emancipated, and as the product of the linear march of progress'. The work of mainstream market environmentalists like Gore, however, closes down any potential questioning of the hegemonic narrative: instead it 'reaffirms *who we are* within the logic of our own civilizational narratives...[and] tells us that capitalism and techno-science, the very tools of our manufactured dystopia, are just as important and legitimate as ever' (2013: 108, original emphasis). As far as Ervine (2013: 93) is concerned, this is a radical political struggle, because she assumes that 'the emergence of modern-day environmentalism represented the ultimate embodiment of Western society's countercultural desires, a radical divergence from the everyday'. Environmental crises present 'existential challenges to capitalism itself', she says, but where the 'logic of traditional environmentalism' took this as the starting point for envisaging 'a new, green world order', figures such as Gore offer a 'path to green deliverance' that simply 'reaffirms our civilizational narratives, while assuaging our guilt' (2013: 110, 93).

Ervine (2013: 96) also develops this idea in explicitly therapeutic terms, arguing that in the face of climate change, 'Westerners...have begun to lose their "self-confidence"' and are suffering from an 'ecologically induced identity crisis'. Noting a 2008 study showing that 'increased personal informedness regarding global warming led to a decrease in concern for the problem, coupled with a sense of less rather than more personal responsibility', Ervine (2013: 96–7) turns to psychoanalysis for answers to this puzzle. She diagnoses denial, dissociation and repression as 'defensive responses to the painful reality, allowing us to maintain the illusion of normalcy despite the mounting contradictory evidence' (2013: 97). Experiencing 'intense feelings of powerlessness, anxiety, loneliness, isolation, and alienation', people turn to 'powerful fantasies that offer simple solutions to complex problems' (2013: 97). Carbon offsetting is one such simple solution, functioning as what Ervine calls a 'psycho-social device' that allows us to contain our anxieties; acting as 'a signifier for *who it is we want to be*' (2013: 108–9, original emphasis). Ervine suggests that if the 'fears and discontents of individuals' were to 'infect the collective societal body', then 'the seeds of counter-hegemony might be sown' so long as these feelings are influenced by 'environmental consciousness' (2013: 98–9). Yet at present those fears and discontents are largely being quietened and contained by 'reassurances that our foundational narratives continue to hold true' (2013: 99).

In other words, we have a collective patient with psychological symptoms which are currently receiving the wrong treatment. Our core beliefs are being shaken by climate change ('phenomena such as global warming challenge the belief that we can ultimately master nature'), yet Gore and other mainstream greens offer misleading 'reassurances that our existence remains stable and that all is well or at least easily corrected' (2013: 103). Instead what is required is 'a fundamental redefinition of modernity that accepts that "mastery is impossible'" (Ervine 2013: 112, quoting Latour). If we can indeed see figures like Gore as providing a comforting story to shore up the self-confidence of modern Western societies, we could equally see Klein (2014: 63) as offering exactly the therapeutic intervention that Ervine recommends, seeking to challenge 'the stories on which Western cultures are founded (that we stand apart from nature and can outsmart its limits)'. Indeed, as we saw earlier, Klein's (2014: 3) self-diagnosis at the beginning of *This Changes Everything* is precisely that she has been in denial about climate change. There is indeed an alternative on offer in Klein's work, but in Ervine's terms we should perhaps understand it more as an alternative treatment plan than a political programme – rejecting techno-science in favour of natural remedies, giving up modernist ambitions to 'mastery' in order to recover older sources of wisdom in 'Indigenous cosmologies' (Klein 2014: 159).

Klein's presentation of these ideas has the form of a leftist political argument, with frequent sharp criticisms of 'unregulated capitalism' and so on, but in content *This Changes Everything* is remarkably similar to DiCaprio's *The 11th Hour*. There, as he sets out the issues in general terms in his pieces to camera, DiCaprio notes that 'growth continues to be the focus of many corporations and governments who deplete our environment for economic gain', and that 'healing the damage of industrial civilization is the task of *our* generation'; a task that requires 'the conscious evolution of our species'. The detail is then filled in by the different contributors interviewed for the documentary, who elaborate and expand on its core themes.

As in Klein's work, one prominent idea in *The 11th Hour* is that it is a mistake to see humanity as separate from nature. According to Kenny Ausubel, founder of the Bioneers organisation, this is the 'most fundamental misunderstanding in the world', because 'the reality is that we are part of nature; in fact, we are nature'. *Truthout* author Thom Hartmann challenges the cultural assumption that 'we are the superior life form on Earth; that we are separate from all other life forms; that we have been given dominion over all other life forms'. Psychologist James Hillman diagnoses the view 'that we're separated from nature' as a 'thinking disorder'. Like other contributors to the film, Hartmann suggests that the cause of this disorder is industrial development based on fossilfuels, noting that while for 'the vast majority of human history humans lived on current sunlight', our problems began when 'we began discovering that there were pockets of ancient sunlight and finding coal here and a little bit of oil there'. Similarly, Brock Dolman, director of the WATER Institute, traces current problems back to when humanity began 'feeding off of the fossil-fuel cycle…a death-based cycle…of dependency on extraction'. Nathan Gardels, the editor of *New Perspectives Quarterly*, also blames 'the industrial revolution', seeing it as a 'great rupture from earlier forms and rhythms of life which were generally regenerative'. When 'nature was turned into a resource that was considered endlessly abundant', he says, this gave rise to 'the idea and the conception behind progress which is limitless growth, limitless expansion'.

As in *This Changes Everything*, contributors to *The 11th Hour* do not hesitate to blame corporations as the contemporary embodiment of this problem or to sympathise with those in what Klein calls the 'sacrifice zones'. Michel Gelobter, then president of the think-tank Redefining Progress, argues that political leaders are only 'responsive to wealth, to money and to corporate power', for example; while Ausubel describes corporations as 'the dominant institution of our age', suggesting that 'the greatest weapon of mass destruction is corporate economic globalization'. Tim Carmichael, president of the Coalition for Clean Air, notes that 'low-income communities of colour' are often the worst affected by pollution; while Omar Freilla, founder of Green Worker Cooperatives, describes such communities as 'dumping grounds'. Sandra Postel, director of the Global Water Policy Project, argues that the response has to be more than 'a matter of tweaking a policy here and there'. What is required is 'a very, very broad societal mobilization' and a 'welling up of involvement of citizens'. Philosophically, former Soviet leader Mikhail Gorbachev suggests that we should reject the mistaken idea that 'man is the king of nature'; while environmental campaigner David Suzuki argues that we need to recover forgotten 'ancient truths'. In that spirit, the film gives the final word to Oren Lyons, Faithkeeper of the Onondaga Nation, who makes the point that 'the Earth has all the time in the world and we don't' – an ending not dissimilar to *This Changes Everything*, which closes with a quote from the Native American environmentalist Henry Red Cloud, to the effect that we need to 'run like a buffalo' in response to the challenge of climate change (see also Klein 2014: 24).

In a sense these similarities may not be so surprising: Klein's film puts a more colourfully anti-capitalist wrapping around what are fairly standard ecological arguments. Perhaps more surprising, for a self-described 'secular Jewish feminist' (Klein 2015), is that she also shares DiCaprio's admiration for the Pope's ecological message. As we saw in Chapter 4, in *Before the Flood* DiCaprio echoes the Pope's call for urgent action to repair 'our common home', even repeating the Holy Father's words as if they were his own. For Klein, invited to the Vatican for the 2015 launch of the environmental encyclical, the dynamic plays out in reverse, as Pope Francis appears to champion her anti-capitalist take on climate change. On Klein's (2015) reading, the encyclical is an exploration of 'the culture of late capitalism', arguing for 'deep changes to our growth-driven economic model', and standing up for 'the trampled victims of a highly unequal and unjust economic system'. While the Pope seems already to be on-board with Klein's radical anti-capitalism, she is conscious of the fact that it is not straightforward to 'reconcile a Christian God with a mystical Earth'. Yet she finds that in some parts of the world 'the more anti-nature elements of Christian doctrine never entirely took hold', and that there the teachings of the Catholic Church are already 'fused' with 'indigenous world views' and 'cosmologies…centered on a living and sacred Earth' (Klein 2015).

The thing that allows this unlikely reconciliation, and the thing that Klein finds most appealing in the Pope's teaching, is his point that 'the Bible has no place for a tyrannical anthropocentrism' (quoted in Klein 2015). In her Vatican remarks, she identifies the questioning of human 'mastery' over nature as the 'core message' at 'the heart of the encyclical':

'We are not God,' the encyclical states. All humans once knew this. But about 400 years ago, dizzying scientific breakthroughs made it seem to some that humans were on the verge of knowing everything there was to know about the Earth, and would therefore be nature's 'masters and possessors,' as René Descartes so memorably put it.[8]

An economy based on fossil fuels allowed that idea to be sustained for a long time, she argues, but 'now we are confronted with the reality that we were never the master' and can only 'save ourselves' if we 'let go of the myth of dominance and mastery'. A similar interpretation is offered by US bishop Father Robert Barron, who understands contemporary environmental problems as flowing from the idea of 'the alienated Cartesian subject going about his work of mastering nature'. The Pope understands, Barron argues, that 'the nature we have attempted to dominate… has now turned on us, like Frankenstein's monster'.[9]

This is the point of agreement: that modernist human hubris is doomed to fail and we should let go rather than attempting to control the natural world. Remarkably, it appears to unite ecologists, Roman Catholics, anti-capitalists, Hollywood celebrities, proponents of Indigenous cosmologies, and contemporary social theorists such as Latour (2003: 36), who all argue that 'mastery is impossible'.

Conclusion: the thrill of letting go

We might understand environmentalist celebrities as part of Kotkin's Clerisy, supporting and legitimising a shift from one declining sector of the capitalist economy to other, rising sectors. Yet more than this, celebrity advocacy helps to lend an appearance of popular legitimacy to elite ideas and activities, producing a simulation of public engagement. In this sense, as Brockington and others counter-intuitively suggest, far from being a means to engage the public, celebrity advocacy helps to disengage and alienate people. Celebrity advocacy enacts, reinforces and legitimises a 'theatrical' politics of performance and spectatorship. As we saw in Chapter 1, this simulation of a political relationship arises from the void of political meaning that followed the collapse of modernist politics at the end of the Cold War. In this context, versions of climate advocacy like Klein's which aim to be more radical end up being more anti-modernist.

It is interesting that both for Klein and for Gore climate change acts as an external imperative for change. It is reminiscent of the way that, as we saw in Chapter 3, Peter Berglez and Ulrika Olausson (2014: 61) understand climate science to work ideologically as 'an abstract authoritative voice', conveying indisputable truths and commanding changes in our everyday lives. It is also interesting that both Klein and Gore appear to derive a certain satisfaction from this. When Gore (2006: 11) speaks of 'the thrill of being forced by circumstances', he seems to find relief or even pleasure in the fact that the spiritual meaning and mission he seeks is not self-motivated but forced upon him by the Truth of climate science which, in this respect at least, turns out to be rather convenient. Where Gore finds that climate change hands him a 'compelling moral purpose' (and perhaps some lucrative opportunities for carbon-trading), Klein (2013) discovers that 'science is telling us all to revolt', and that the 'revolutionary nature of climate science' provides 'the best argument we have ever had' for changing the 'rules of capitalism'.[10] As we saw earlier, for Klein (2011) climate change is not only a 'catalyst' for social change but also coheres a 'progressive' agenda around a 'clear scientific imperative'. Klein and Gore, revolutionary radical and establishment centrist, are politically different of course, but both seem to think that their own position is supported and given extra weight and validity by the external imperative of climate change.

Perhaps that 'thrill of being forced by circumstances' is part of the pleasure of giving up pretentions to 'mastery'. It does, however, have the unfortunate side-effect that it also undermines the foundational ideas of modernist political subjectivity and seems inexorably to converge, even in radical versions like Klein's, with a therapeutic discourse of self-help and personal healing.

Notes

1 *Climate Change & Influential Spokespeople: A Global Nielsen Online Survey*. Oxford: The Nielsen Company / Oxford University Environmental Change Institute, June 2007.
2 *Attitudes to Climate Change – Amongst Young People – Wave 2, Key Research Findings*. Central Office of Information, May 2008, COI ref. 286957. Available at: http://web.archive.org/web/20081207022514/http://www.defra.gov.uk/environment/climatechange/uk/individual/attitudes/pdf/cc-youth-tracker-presentation.pdf.
3 See the reports of these and other events on the Leonardo DiCaprio Foundation website: http://leonardodicaprio.org/news/.
4 *The Big Idea*, BBC2, 14 February 1996, available at https://www.youtube.com/watch?v=GjENnyQupow, with a transcript at http://citystrolls.com/articles/the-necessary-illusion/. Chomsky was drawing here on the ideas of US political scientist Thomas Ferguson.
5 Page's email to Clinton's campaign team was one of those released by Wikileaks in 2016. See: https://wikileaks.org/podesta-emails/emailid/1352. The Clinton campaign video in question is available at: https://www.youtube.com/watch?v=_ZwguLJVxsM.
6 See Klein 2014: 136–8. According to the eco-modernists at Environmental Progress, 'German emissions increased in 2016 for a second year in a row as a result of the country closing one of its nuclear plants and replacing it with coal and natural gas....Not only did new solar and wind not make up for the lost nuclear, the percentage of time during 2016 that solar and wind produced electricity declined dramatically'. See: German Emissions Increase in 2016 Due to Nuclear Plant Closure, *Environmental Progress*, 13 January 2017, www.environmentalprogress.org/big-news/2017/1/13/breaking-german-emissions-increase-in-2016-for-second-year-in-a-row-due-to-nuclear-closure.
7 The *Guardian* film, *Under the Surface*, is available at: www.theguardian.com/environment/video/2016/nov/07/naomi-klein-at-the-great-barrier-reef-under-the-surface.
8 Naomi Klein's Press Statement at the Vatican, *The Leap*, 1 July 2015, https://theleapblog.org/watch-naomis-press-statement-at-the-vatican/.
9 See Fr. Robert Barron, Laudato Si and Romano Guardini, *Word on Fire* (no date) https://laudatosi.com/watch.
10 This article appeared in the 23 October 2013 issue of the *New Statesman*, guest-edited by comedian and actor Russell Brand, which included several celebrity contributions. The issue's theme was 'Revolution of Consciousness'. See http://www.newstatesman.com/staggers/2013/10/weeks-new-statesman-russell-brand-guest-edit.

References

Anderson, Alison (2011) Sources, media, and modes of climate change communication: The role of celebrities, *WIREs Climate Change*, 2 (4): 535–46.

Berglez, Peter and Ulrika Olausson (2014) The post-political condition of climate change: An ideology approach, *Capitalism Nature Socialism*, 25 (1): 54–71.

Boykoff, Maxwell T. and Michael K. Goodman (2009) Conspicuous redemption? Reflections on the promises and perils of the 'celebritization' of climate change, *Geoforum*, 40 (3): 395–406.

Brockington, Dan (2008) Powerful environmentalisms: Conservation, celebrity and capitalism, *Media, Culture & Society*, 30 (4): 551–68.

Brockington, Dan (2014) *Celebrity Advocacy and International Development*. Abingdon: Routledge.

Brockington, Dan (2016) Epilogue: The politics of celebrity humanitarianism, in Lisa Ann Richey (ed.) *Celebrity Humanitarianism and North–South Relations*. Abingdon: Routledge.

Brockington, Dan and Spensor Henson (2014) Signifying the public: Celebrity advocacy and post-democratic politics, *International Journal of Cultural Studies*, 13 (2): 107–26. DOI:10.1177/1367877914528532 [accessed electronically].

Chouliaraki, Lilie (2012) The theatricality of humanitarianism: A critique of celebrity advocacy, *Communication and Critical/Cultural Studies*, 9 (1): 1–21. DOI:10.1080/1479142 0.2011.637055 [accessed electronically].

Couldry, Nick and Tim Markham (2007) Celebrity culture and public connection: Bridge or chasm? *International Journal of Cultural Studies*, 10 (4): 403–21.

Crouch, Colin (2004) *Post-Democracy*. Cambridge: Polity.

Doyle, Julie, Nathan Farrell and Michael K. Goodman (forthcoming) Celebrities and Climate Change: History, Politics and the Promise of Emotional Witness, in Matthew C. Nisbet (ed.) *The Oxford Encyclopedia of Climate Change Communication*. Oxford: Oxford University Press, http://climatescience.oxfordre.com/page/climate-change-communication/.

Ervine, Kate (2013) Al Gore as Carbon Warrior: The Politics of Inaction, in Gavin Fridell and Martijn Konings (eds.) *Age of Icons: Exploring Philanthrocapitalism in the Contemporary World*. Toronto: University of Toronto Press.

Ervine, Kate (2015) This changes everything: Capitalism vs. the climate, by Naomi Klein (book review), *Canadian Journal of Development Studies*, 36 (3): 416–18.

Furedi, Frank (2010) Celebrity Culture, *Society*, 47 (6): 493–97.

Goodman, Michael K. (forthcoming) Environmental Celebrity, in Noel Castree, Mike Hulme and Jim Proctor (eds.) *The Routledge Companion to Environmental Studies*. London: Routledge.

Goodman, Michael K., Jo Littler, Dan Brockington and Maxwell Boykoff (2016) Spectacular environmentalisms: Media, knowledge and the framing of ecological politics, *Environmental Communication*, 10 (6): 677–88.

Gore, Al (2006) *An Inconvenient Truth*. London: Bloomsbury.

Gunster, Shane (2017) This changes everything: Capitalism vs. the climate (book review), *Environmental Communication*, 11 (1): 136–8.

Hibberd, Matthew and An Nguyen (2013) Climate change communications and young people in the Kingdom: A reception study, *International Journal of Media and Cultural Politics*, 9 (1): 27–46.

Hirsch, Jerry (2015) Elon Musk's growing empire is fueled by $4.9 billion in government subsidies, *Los Angeles Times*, 30 May, www.latimes.com/business/la-fi-hy-musk-subsidies-20150531-story.html.

Kapoor, Ilan (2013) *Celebrity Humanitarianism*. Abingdon: Routledge.

Klein, Naomi (2011) Capitalism vs. the climate, *The Nation*, 9 November, www.thenation. com/article/capitalism-vs-climate/.

Klein, Naomi (2013) How science is telling us all to revolt, *New Statesman*, 29 October, www.newstatesman.com/2013/10/science-says-revolt.

Klein, Naomi (2014) *This Changes Everything: Capitalism vs. the Climate*. London: Penguin.

Klein, Naomi (2015) A radical Vatican? *New Yorker*, 10 July, www.newyorker.com/news/news-desk/a-visit-to-the-vatican.

Klein, Naomi (2016) Climate change is intergenerational theft. That's why my son is part of this story, *Guardian*, 7 November, www.theguardian.com/environment/2016/nov/07/climate-change-is-intergenerational-theft-thats-why-my-son-is-part-of-this-story.

Kotkin, Joel (2014) *The New Class Conflict*. Candor, NY: Telos Press.

Latour, Bruno (2003) Is re-modernization occurring – and if so, how to prove it? *Theory, Culture & Society*, 20 (2): 35–48.

Littler, Jo (2008) 'I feel your pain': Cosmopolitan charity and the public fashioning of the celebrity soul, *Social Semiotics*, 18 (2): 237–51.

Markham, Tim (2015) Celebrity advocacy and public engagement: The divergent uses of celebrity, *International Journal of Cultural Studies*, 18 (4): 467–80.

Meyer, David S. and Joshua Gamson (1995) The challenge of cultural elites: Celebrities and social movements, *Sociological Inquiry*, 65 (2): 181–206.

Moran, Jose (2017) Time for Tesla to listen, *Medium*, 9 February, https://medium.com/@moran2017j/time-for-tesla-to-listen-ab5c6259fc88#.lot6055yc.

Nunn, Heather and Anita Biressi (2010) 'A trust betrayed': Celebrity and the work of emotion, *Celebrity Studies*, 1 (1): 49–64.

Richards, Barry (2004) The emotional deficit in political communication, *Political Communication*, 21 (3): 339–52.

Richards, Barry (2007) *Emotional Governance*. Basingstoke: Palgrave Macmillan.

Rojek, Chris (2012) *Fame Attack: The Inflation of Celebrity and Its Consequences*. London: Bloomsbury.

Rojek, Chris (2014) 'Big Citizen' celanthropy and its discontents, *International Journal of Cultural Studies*, 17 (2): 127–41.

Thrall, A. Trevor, Jaime Lollio-Fakhreddine, Jon Berent, Lana Donnelly, Wes Herrin, Zachary Paquette, Rebecca Wenglinski and Amy Wyatt (2008) Star power: Celebrity advocacy and the evolution of the public sphere, *The International Journal of Press/Politics*, 13 (4): 362–85.

Wong, Julia Carrie (2017) Tesla factory workers reveal pain, injury and stress: 'Everything feels like the future but us', *Guardian*, 18 May, www.theguardian.com/technology/2017/may/18/tesla-workers-factory-conditions-elon-musk.

CONCLUSION

In search of the political

> Climate reductionism…is nurtured by elements of a Western cultural pessimism which promote the pathologies of vulnerability, fatalism and fear….By handing the future over to inexorable non-human powers, climate reductionism offers a rationalisation, even if a poor one, of the West's loss of confidence in the future.
>
> (Hulme 2011: 265–6)

If, as Chris Methmann (2014: 1) suggests, the 2009 Copenhagen climate summit was a crystallising moment for post-political approaches to climate change, it also prompted both academics and activists to challenge the post-political. Below we consider that challenge, the main thrust of which has been to link climate change to the wider problems of contemporary capitalism. Yet as we saw in Chapter 5, using 'capitalism' or 'neoliberalism' as boo-words is not in itself a hugely radical thing to do today, in the sense that it can be woven into an inward-looking therapeutic discourse of self-healing rather than necessarily directing attention outward to changing external circumstances. What really matters in this respect is how we understand political subjectivity, but this is a topic on which critics of post-political consensus can sometimes appear evasive. Moreover, as was suggested in our discussion of 'ethical' or 'green' consumerism in Chapter 3, and of 'commodified' forms of celebrity activism in Chapter 4, an anti-capitalist critique often seems to miss its intended target. At least in some versions, what presents itself as a 'critical' perspective instead turns out to be close to the dominant, mainstream outlook. Perhaps that is not so surprising since, as was argued in Chapters 1 and 2, many inherited assumptions about where to draw the boundary between the mainstream and the critique offer an unreliable guide to the present.

Activists against the post-political

A number of scholar-activists have recounted their experiences of protests that explicitly set out to challenge the post-political framing of climate change. Anneleen Kenis and Erik Mathijs (2014: 148) analyse the Climate Justice Action network as 'one of the most prominent movements in recent history that explicitly took issue with the consensual, post-political logic governing much of the debate on climate change'. David Featherstone (2013) and Paul Chatterton *et al.* (2013) reflect on their involvement with debates at the Klimaforum, the 'alternative climate summit' at Copenhagen in 2009, and with other subsequent initiatives. Andrew Bowman (2010) assesses the limitations of activist groups such as Plane Stupid, Rising Tide, Climate Rush and Camp for Climate Action in seeking to politicise the issue. What is clear from these accounts is that at least some activist organisations have not only recognised the problem of the post-political, but have also either sought to avoid it or have made it into the target of their protests and campaigns.

Particularly striking in this respect is Kenis and Mathijs's (2014: 149) account of the Climate Justice Action movement (CJA), which 'did not merely advocate a specific cause' but also 'targeted post-politics as an obstacle for promoting this cause', engaging in 'a type of meta-struggle for genuine political struggle and disagreement to even become possible and visible'. In its campaign literature and activities at Copenhagen in 2009, the CJA 'aimed at "breaking" the consensus ideology of the climate summit and turning the spotlight to what was neglected in the hegemonic discourse' (2014: 152). This meant ignoring or explicitly dismissing many of the points highlighted in the 'hegemonic climate discourse', adopting slogans like 'it is not just about CO_2', 'forget shorter showers', or 'climate change is not an environmental issue'; and instead arguing that 'climate change is a symptom, capitalism the crisis' (Kenis and Mathijs 2014: 151–2). Similarly, Bowman (2010: 173–4) describes how 'radical climate activists have posited the climate crisis as an opportunity to create a more just and equitable future society', adopting slogans such as 'social change not climate change' and 'government and corporations are not the answer'.

According to some, radical activists have successfully escaped the post-political trap. For Featherstone (2013: 45), 'vociferous demands for "climate justice"' already provide a 'distinctive approach to the politics of climate change', and one that presents a 'significant challenge to the dominant terms of climate change politics'. Accounts of the post-political may accurately describe the mainstream, he argues, but 'it is unhelpful to extrapolate from this that all climate change politics is necessarily de-politicizing'. Overemphasising the problem of the post-political, in other words, ignores the 'important ways in which dominant responses to climate change are being brought into contestation' (2013: 49). From this perspective, there is no intellectual problem at all, in fact: in terms of its ideas, radical, oppositional politics carries on just as well as ever. The problem instead is understood as a purely practical one: activists tend to get shut out or arrested. It is this 'dramatic repression of peaceful protest' that is the real issue (2013: 57). Similarly, arguing that 'climate justice involves an antagonistic framing of climate politics that breaks with attempts to construct climate change as a "post-political" issue', Chatterton *et al.* (2013: 602, 618) suggest that any '"post-political" consensus is an active process achieved through the disciplining work of repressive policing and juridical frameworks'.

This reduction of the problem does not quite ring true, for two reasons. First, these critics protest too much: if there really was no problem (other than evading arrest at demonstrations), then it seems a peculiar waste of time to sit down and write articles complaining about the exaggeration of the 'post-political'. Other people think radical politics is weak, but you know it to be strong: would you not simply prove them wrong by getting on with it? Second, it seems clear from subsequent experience that the problem of the post-political has not gone away. The 2014 UN Climate Summit may have been disappointingly consensual, but out on the streets of New York the People's Climate March offered 'a glimpse of a far more urgent, motivated climate justice movement', claimed Naomi Klein (in Darby 2014). Yet according to the left-wing muckrakers at *Counterpunch*, the march was a 'corporate PR campaign' with 'no demands, no targets, and no enemy' (Gupta 2014). It also had 'no unity other than money': 'One veteran activist…said it was made known there was plenty of money to hire her and others'. The PR and marketing ethos shaping the event 'turned social justice into a product to enhance the liberal do-gooding lifestyle'. Radicals getting arrested on the sidelines of such events does not so much offer an alternative as lend a little extra credibility to the post-political simulacrum. As Bowman (2010: 175) points out, 'Despite the tactical illegality and flirtation with revolutionary rhetoric, radical climate politics enjoys a high level of elite support compared to other folk devils of radical protest'. We should not confuse the 'radical, oppositional tactic of direct action' with 'radical, oppositional politics', he argues. Even rowdy, law-breaking and ostensibly radical protests often 'operate more as a means of communicating with elites…"sending a message to the government" that they have been insufficiently technocratic, having failed in their carbon counting, and failed to [rein] in greedy consumers' (Bowman 2010: 185).

Kenis and Mathijs (2014: 153) provide a frank assessment of the problems of tackling the post-political. A key weakness of the CJA network, they acknowledge, is that while activists were clear that they were fighting for systemic change, they were not able to articulate clearly what this meant:

> First, to the extent that they existed, visions of alternatives were rather vague. Several activists admitted that they did not know what an alternative would look like….Second, to the extent activists were able to formulate visions of alternatives, they diverged widely: from living differently and more soberly in one's personal life to a democratically planned economy, from political reform to a change of mentality, from communal living and cooperative pro-duction to a stronger state.
>
> (Kenis and Mathijs 2014: 153)

Of course, as one of their interviewees pointed out, it would be unfair to expect them to all come up with the same, well worked-out blueprint of the ideal alternative society. Yet there does appear to be a certain incoherence and vagueness to notions of 'system change' and 'climate justice'. The activists interviewed by Kenis and Mathijs did clearly identify the problem as 'neoliberalism and capitalism' (2014: 154), but, as discussed further below, it is by no means certain that simply using these terms really clarifies matters.

A common concern that emerges in the work of these authors is that the urgency of climate threat is used as a reason to shut down political debate. Featherstone (2013: 52) for example, complains that the prominent environmentalist Mark Lynas has 'argued that...climate change is such an urgent problem that struggles for equity need to take second place' (see Lynas 2010). Similarly, Bowman (2010: 183) recounts with evident disgust how the *Guardian*'s George Monbiot was 'treated with almost messianic levels of reverence' by climate camp activists, despite his pragmatic insistence that 'challenges to capitalism and state-corporate power' must not be allowed to interfere with the main priority of 'meeting emissions targets' (see Monbiot 2008). Kenis and Mathijs (2014: 153) note that even the CJA activists, who tried to resist the depoliticising pressure of 'the clock ticking', were sometimes tempted into adopting an 'apocalyptic vision'. As Bowman (2010: 187) remarks in a sardonic paraphrase of Fredric Jameson, 'In the time-frame the science prescribes, it truly is easier to imagine the end of the world than the end of capitalism'. There is a logic to this pragmatism, of course, if one believes that 'the apocalypse now has a fairly precise date in the very near future...[in] 10 years, 100 months, the point at which we reach 2 degrees of warming' (Bowman 2010: 187). If the catastrophic and irreversible 'tipping point' is only a few years or even months away, then taking time out to build a global movement for 'systemic change' might look like a dangerous distraction.

Yet perhaps there is another way to read this anti-political apocalypticism: like invocations of The Science, it provides the 'thrill of being forced by circumstances' that we discussed at the end of Chapter 5. That is to say, rather than a closing down of lively radical politics, it offers a substitute for absent political argument, an irresistible imperative to change that does not have to be fought and argued for. Hence when called on for political justification, activists will say things like Plane Stupid's co-founder Joss Garman: 'this isn't about ideals so much as hard science' (quoted in Bowman 2010: 182).[1] 'This Changes Everything', it turns out, can be another way of saying 'There is No Alternative'.

Subjects, citizens and consumers

Claims that left-wing politics is alive and well but just out of sight (or spending the night in a police cell) are unconvincing. A theme that runs throughout this book is that critique often seems to misidentify the target today. This is particularly clear in relation to the critique of the 'commodified' solutions and 'heroic' individualism that are allegedly presented in the media and celebrity culture. The critique is presented as a radical analysis of neoliberal governance, but in fact entails joining in with an already dominant antipathy to modernist conceptions of political subjectivity. As we saw in Chapter 3, for most analysts of ethical consumerism the key thing is to remove or minimise the negative 'consumerism' part and focus instead on the positive 'ethical deliberation' aspect, thereby breaking any association between ethically 'responsibilised' subjects and neoliberal ideas of consumer sovereignty. What is rejected, however, is not simply *consumer* sovereignty, but the notion of the sovereign subject as such.

In questioning 'whether neo-liberalism is a coherent, *ambitious* programme of rule; and whether it [aims] to extend itself by bringing into existence fully-formed neo-liberal subjects', for example, Clive Barnett *et al.* (2008: 626, original emphasis) offer two – rather contradictory – grounds for concluding that it does not. On one hand, although neoliberal governance may entail attempts to 'govern consumption', this does not necessarily mean 'governing people's identities as "consumers"' (2008: 633–4). Ethical consumer campaigns, they argue, do not involve the strategic achievement of 'strong interpellative subject-effects', but leave room for 'lay normativity': space for people to negotiate with, reflect upon, and interpret 'moral imperatives to consume more "ethically", "responsibly", or "sustainably"' (2008: 636, 647–8, 644). On the other hand, however, Barnett *et al.* also make the argument against 'strong interpellative subject-effects' on the grounds that, at least some of the time, ethical consumer campaigns involve making 'changes to consumption practices that go on "behind people's backs"' – for instance by ensuring that the only available choice in a particular location is the ethical, sustainable or fair trade option – and therefore do not 'require direct interventions to regulate individual consumer choice at all' (2008: 637). Clearly, an intervention to shape the 'choice architecture' in order to produce the required consumer behaviours does not need to 'interpellate' people ideologically, but neither does it seem to give much weight to the 'lay normativity' which is supposedly the seedbed for creative ethical and political engagement.

Barnett *et al.* attempt to square this circle in two ways. First, they emphasise that 'Consumer choice is best understood…in terms of the embedding of subjectively motivated action in patterns of practice', so that the pre-shaping of those patterns appears to be less of a contradiction. Second, they argue that successful consumer campaigns involve a participatory ethos of 'shared learning through peer groups and social networks', for example by 'identifying and recruiting "intermediaries" in peer networks who persuade and influence others in conversation' (2008: 644, 643; and see further Barnet 2010). It is important to note that the sort of thing they have in mind here (they cite work by 'UK think-tanks such as the Green Alliance and Demos') is actually concerned with official communications initiatives, rather than bottom-up, consumer-led campaigning. Demos and the Green Alliance produced a report for the UK Department for Environment, Food and Rural Affairs on how 'policymakers can co-opt the techniques used so effectively in the marketing of consumer goods to achieve environmental and social, alongside commercial goals' (Collins *et al.* 2003: 8). There is a notion of participation here, but it seems to be mostly about meeting the needs of elites rather than fostering some sort of grassroots movement for social and political change.

It may be that Barnett *et al.* are led to make somewhat contradictory arguments about ethical consumer campaigns because they are describing a mode of governance which is indeed contradictory, in ways that possibly look odd when considered in the abstract, but which make sense if seen in historical context. The issues they discuss in terms of theoretical and methodological considerations (Barnett *et al.* 2008: 625) are better understood as the product of a particular set of historical circumstances. In the 1980s, 'neoliberal' governments did have something more like a 'coherent, ambitious programme of rule', albeit one largely defined negatively against the targets of organised labour and state welfarism. With the collapse of modern Left/Right politics at the end of the decade, however, it became much more difficult to imbue neoliberal rationalities of governance with meaningful political content. In the UK, for example, the main innovation of John Major's Conservative government was the launch in 1991 of the 'Citizen's Charter', which explicitly articulated, and institutionalised, an understanding of the citizen as a customer or consumer of services such as education or health provision (Drewry 2005). Yet compared with the previous Thatcher administration, this represented a turn away from political ideology toward a technocratic, managerial approach to policy, addressing the individual as a consumer-citizen but not as part of any clear political project. As Gavin Drewry (2005: 325) observes, the Charter remained a 'core part of the Conservative government's programme' until the 1997 election, when it was then 'repackaged and relaunched by Tony Blair's Labour administration'. In the 1990s, in other words, addressing citizens as consumers became entrenched as a way to rationalise and justify governance without needing to articulate any political grand narrative or to produce the sort of 'strong interpellative subject-effects' which Barnett *et al.* note are absent.

This, then, is the context for the incorporation of 'ethical consumerism' into government policy initiatives: the post-Cold War crisis of meaning discussed in Chapter 1. The Demos/Green Alliance report cited above, for instance, basically offers advice on 'how governments should cope with situations of uncertainty and declining public confidence' (Collins *et al.* 2003: 10). Beginning from the premise that 'a trust deficit now cuts across almost every aspect of contemporary British society' and noting that the public have 'an overall impression that government cannot be trusted', it quickly establishes that an 'expert-led, command-and-control approach to public influencing', as exemplified in wartime propaganda campaigns demanding 'personal sacrifice or behaviour-change, for the sake of the greater good', is not tenable in today's 'complex, diverse and individualised society' where 'a simple, linear theory collides with a more complex reality' (2003: 8–14).[2] The Demos/Green Alliance writers recommend more 'open and participatory models of communications', which work in a more informal and 'viral' way, using the 'intermediaries or "network hubs" able to influence others to change behaviour' (2003: 10, 16–17). Similarly, a 2009 report by the Institute for Public Policy Research found a 'general cynicism about the motivations of the Government in pushing for action on climate change', but suggested that this could be overcome by adopting the more indirect approach of mobilising the peer-to-peer influence of so-called 'Now People': trend-setters who are 'particularly powerful…when it comes to determining consumption-related behaviours' (Platt and Retallack 2009: 4–5). Such thinking puts the notion of engaging people's 'lay normativity' as 'ethical consumers' in a rather different light.

The new relationship that started to develop between governments and citizen-consumers was not grounded in strong political identifications, but it still required legitimacy (Varul and Wilson-Kovacs 2008: 3), at the very least in order to provide elite self-justification, and it required some level of wider participation. In casting that participation in term of ethical reflexivity, Barnett and others are really describing, not so much a political relationship between the state and citizen-consumers, but a therapeutic mode of governance. For Barnett *et al.* (2008: 641), this should quite literally be understood in therapeutic terms – in explaining the 'ethical problematization' of the consumer they draw on Ian Hodges's (2002) account of 'the ethical operation of therapeutic discourse' in psychological counselling. More broadly, the relationship between citizen-consumers and the enabling state that is enacted in ethical consumerism can be understood as a therapeutic one because it is primarily concerned with inward-focused work on the self. As Barnett *et al.* (2008: 649, 641) explain, in ethical consumption people are presented with 'dilemmas and conundrums' and are 'encouraged and empowered to problematize their own conduct, to make a "project" out of various aspects of their lives'.

Similarly, Matthias Varul and Dana Wilson-Kovacs (2008: 11, 2) note that although 'ethical selving' often involves virtue signalling and displays of good taste, ethical consumption is appealing to those with 'an inclination towards scrutinising, questioning, weighing the consequences of [their] behaviour', since 'the essential aspect seems to be self-reassurance about being a morally acceptable person'. Their study of fair-trade consumers also found that:

> Participants inter-weaved accounts of what fairtrade goods they buy with accounts of who they are (often accompanied by a biographical account) in a way that such goods are to be regarded [as] psychologically significant 'symbols of the self'…or 'extensions of the self'…that are clearly invested with identity, asserting the self as an *ethical* self.
>
> (Varul and Wilson-Kovacs 2008: 7, original emphasis)

Barnett *et al.* (2008: 645) are no doubt correct to argue that the 'responsibilization of everyday life' cannot be understood in terms of '"vertical" processes of positioning by hegemonic discourses' and instead involves the incorporation of notions such as ethical or sustainable consumption into one's own 'narrative-self', but this is more a therapeutic process of self-reflection than the cultivation of an externally focused political subjectivity.

In terms of the media representations of ethical consumption considered in Chapter 3 – whether depoliticised and impartial, or campaigning and partisan – climate change as ideology can be seen to operate less as an overtly top-down, public and political 'hegemonic discourse' and more as an informal, personal and therapeutic mode of address. Similarly, as discussed in Chapter 4, the point about celebrity climate activism is not that it offers heroic role models, but that it addresses us as vulnerable, 'neurotic' subjects who are engaged in self-monitoring projects of the self. The targets of critique – acquisitive consumerism, bourgeois individualism – are mere straw men. The 'positives' that critics seek to discover in either ethical consumption or celebrity advocacy – the reflexive, ethical self-auditing and the emotional displays of caring – are part of the mechanisms of contemporary governance, not the critical alternatives to it.

Taking the subject out of therapy

As we noted in the Introduction, a therapeutic ethos has become an important feature of contemporary public life and is not something that is limited to green politics, but there does seem to be a particular affinity. Organisations such as The Transition Network, or Carbon Conversations, for example, combine environmentalism with therapeutic support. Carbon Conversations, for instance, is a 'psychosocial project that addresses the practicalities of carbon reduction while taking account of the complex emotions and social pressures that make this difficult'. It offers advice on 'the psychology of climate change' and helps people to 'overcome their fears and defensiveness in dealing with it'.[3] While both of these organisations were established in the mid-2000s, the Work That Reconnects Network traces its history back to the late 1970s, and was initially known as Despair and Empowerment work, offering 'deep ecology' as an antidote to the unhappiness of people 'growing up in the Industrial Growth Society'.[4]

If these ideas have more resonance in media and popular culture today, it is not because ecological therapy has won over more converts, nor because environmentalism necessarily produces 'weak' subjects in need of support and at risk of burnout and despair. Rather, it is symptomatic of a wider problem of political subjectivity that finds explicit expression in ecological approaches which, as we saw in Chapter 5, trace the roots of current problems back to the Cartesian subject and Enlightenment science. Ecology offers a fearful vision of the future, but so do mainstream political leaders. Ecology emphasises limits, distrusts modernist hubris and repudiates anthropocentrism, but these are not marginal ideas today. Ecology has put the political subject into therapy, but we live in a therapeutic culture.

In most discussions of climate change as a 'post-political' issue, there is one aspect of this idea that usually tends to be ignored or glossed over by later writers – namely that in critiquing ecology as a 'new opium of the people', both Slavoj Žižek and Alain Badiou were explicitly seeking to defend the Cartesian subject. For Žižek, an 'ecology of fear' has

> every chance of developing into the predominant form of ideology of global capitalism, a new opium for the masses replacing the declining religion: it takes over the old religion's fundamental function, that of having an unquestionable authority which can impose limits. The lesson this ecology is

constantly hammering away at is our finitude: we are not Cartesian subjects extracted from reality, we are finite beings embedded in a biosphere which vastly transcends our horizon.

(Žižek 2008: 439)

Žižek was agreeing with Badiou, who in the interview he was alluding to here had said: 'I am Cartesian: man is the master and possessor of nature. That has never been as true as today' (in Feltham 2008: 139). Indeed, Žižek had already written a book about the manifold intellectual perspectives which, despite their differences, were all 'united in their rejection of the Cartesian subject' (2000: 2). Representatives of these currents of thought included the 'Deep Ecologist' who 'blames Cartesian mechanicist materialism for providing the philosophical foundation for the ruthless exploitation of nature'. Also included, though, were the 'Habermasian theorist of communication', the 'Heideggerian proponent of the thought of Being', the cognitive scientist, the 'critical (post-) Marxist', the feminist, 'the whole world' (2000: 1–2). Whatever else he may be wrong about, Žižek is surely right about this: in fact, since the turn of the century several more, newly fashionable intellectual currents – post-humanism, new materialism, actor-network theory, object-oriented ontology – have now joined the ranks of those hostile to the Cartesian subject.[5]

These intellectual assaults on modernist understandings of political agency arise from the same dynamic that has given rise to therapy culture and its invitation to understand ourselves as vulnerable, neurotic subjects engaged in reflexive self-monitoring rather than hubristically attempting to shape the object world through outward-facing engagement and action. Ecological perspectives, already close to this outlook, give it powerful expression in relation to climate change. As Monbiot (2009) puts it, for example, the battle against climate change is also 'a battle to redefine humanity': the line is drawn not between Left and Right, but between 'expanders and restrainers'. Erik Swyngedouw (2013: 2) argues that 'if we really care deeply about the climate' we need to shift our 'theoretical gaze and political passions', from 'a concern with the environment per se to a concern and passion for the construction of a different politics'. We could go further – the problem is not so much that environmental problems require a political response, but that focusing on the issue of climate change will not help with challenging the post-political consensus: quite the opposite.

As we have seen in relation to political rhetoric, celebrity campaigning and news media framings of climate change, there are indeed ways in which particular constructions of the issue work to depoliticise it, whether in terms of moral certainty, personal lifestyle choices or consensus agreement. Yet as we have argued, attempts to construct a more radical, politicising appeal – whether through a more properly ecological humility toward nature, a more pronounced anti-consumerism, or a more emotionally charged mode of address – tend to make matters worse. It is simply mistaken to assume that there was once some radical core in environmental politics that can be recaptured or reignited so as to overcome the problem of the post-political. The radicalism of the ecological outlook – its suspicion of technological and scientific progress, its understanding of 'industry' or 'extractivism' as the problem, its hostility to hubristic attempts to master or control nature, and its advocacy of respect for natural limits – is part of the problem, rather than a solution to it. Perhaps the greatest mistake in this respect is the idea that greater emotionalism will provide the answer to the post-political condition. In today's circumstances, such emotional appeals tend to further reinforce a therapeutic outlook that encourages us to understand politics in terms of a project of the self rather than changing the world.

Notes

1   Garman gave a similar answer when interviewed by the BBC's Ethical Man in 2006. Challenged over whether it was not 'very, very arrogant' to disrupt people's holiday flights by airport protests, he replied 'no, we haven't decided that, the science dictates that'. See: A flight that almost cost me my marriage (video), 4 December 2009 (first broadcast October 2006), www.bbc.co.uk/blogs/ethicalman/2009/12/the_tv_stunt_that_almost_cost_me_my_marriage.html

2   The idea that modernist understandings have to give way in the face of an appreciation of non-linearity and complexity applies not only to communications but to governance as such. This is arguably a new development from neoliberal ideas, which involves the incorporation of 'critical' perspectives (Chandler 2014).

3   See http://www.carbonconversations.co.uk/p/about.html and https://transitionnetwork.org/about-the-movement/.

4   See http://workthatreconnects.org/about-network/. I am grateful to Jo Hamilton for drawing my attention to this phenomenon with her paper, *The Emotional Climates of the Everyday*, presented at the Media, Communication and Cultural Studies Association (MeCCSA) Annual Conference, Leeds University, 13 January 2017.

5   Žižek (2008: 461) has offered some proposals on how to recast climate change in more properly political terms, which centre on the idea of 'egalitarian terror'. They comprise 'strict egalitarian justice' (equal norms of per capita energy consumption), terror ('ruthless punishment' of violations), voluntarism (in the form of unspecified 'large-scale collective decisions'), and finally 'trust in the people' (which oddly includes recruiting informers who will denounce carbon criminals to 'the authorities'). Possibly this is just a very bad joke, although Sharpe and Boucher (2010: 192) think that Žižek's ecological examples are 'politically correct'; chosen to 'sound somewhat acceptable in today's world', and to thereby sugar-coat his advocacy of 'egalitarian terror'. For a critique of the latter concept see Johnson (2011).

References

Barnett, Clive (2010) The politics of behaviour change, *Environment and Planning A*, 42 (8): 1881–6.

Barnett, Clive, Nick Clarke, Paul Cloke and Alice Malpass (2008) The elusive subjects of neo-liberalism, *Cultural Studies*, 22 (5): 624–53.

Bowman, Andrew (2010) Are We Armed Only with Peer-Reviewed Science? The Scientization of Politics in the Radical Environmental Movement, in Stefan Skrimshire (ed.) *Future Ethics: Climate Change and Apocalyptic Imagination*. London: Continuum.

Chandler, David (2014) *Resilience: The Governance of Complexity*. Abingdon: Routledge.

Chatterton, Paul, David Featherstone and Paul Routledge (2013) Articulating climate justice in Copenhagen: Antagonism, the commons, and solidarity, *Antipode*, 45 (3), 602–20.

Collins, Joanna, Gillian Thomas, Rebecca Willis and James Wilsdon (2003) *Carrots, Sticks and Sermons: Influencing Public Behaviour for Environmental Goals*. A Demos/Green Alliance report produced for Defra. London: Demos/Green Alliance.

Darby, Megan (2014) Naomi Klein: New York showed glimpse of climate justice movement, *Climate Home*, 27 October, www.climatechangenews.com/2014/10/27/naomi-klein-new-york-showed-glimpse-of-climate-justice-movement/.

Drewry, Gavin (2005) Citizen's charters: Service quality chameleons, *Public Management Review*, 7 (3): 321–40.

Featherstone, David (2013) The contested politics of climate change and the crisis of neo-liberalism, *ACME: An International E-Journal for Critical Geographies*, 12 (1): 44–64.

Feltham, Oliver (2008) *Alain Badiou: Live Theory*. London: Continuum.

Gupta, Arun (2014) How the People's Climate March became a corporate PR campaign, *Counterpunch*, 19 September, www.counterpunch.org/2014/09/19/how-the-peoples-climate-march-became-a-corporate-pr-campaign/.

Hodges, Ian (2002) Moving beyond words: Therapeutic discourse and ethical problematiza-tion, *Discourse Studies*, 4 (4): 455–79.

Hulme, Mike (2011) Reducing the future to climate: A story of climate determinism and reductionism, *Osiris*, 26 (1): 245–66.

Johnson, Alan (2011) The power of nonsense, *Jacobin*, 14 July, www.jacobinmag.com/2011/07/the-power-of-nonsense.

Kenis, Anneleen and Erik Mathijs (2014) Climate change and post-politics: Repoliticizing the present by imagining the future? *Geoforum*, 52: 148–56.

Lynas, Mark (2010) Why it's wrong to preach 'climate justice', *New Statesman*, 14 January, http://www.newstatesman.com/blogs/the-staggers/2010/01/lynas-climate-carbon.

Methmann, Chris (2014) *We are all green now: Hegemony, governmentality and fantasy in the global climate polity* (PhD diss.). Hamburg: Staats- und Universitätsbibliothek Hamburg, http://d-nb.info/1050818407/34.

Monbiot, George (2008) Climate change is not anarchy's football, *Guardian*, 22 August, www.theguardian.com/commentisfree/2008/aug/22/climatechange.kingsnorthclimatecamp.

Monbiot, George (2009) This is bigger than climate change. It is a battle to redefine humanity, *Guardian*, 14 December, www.guardian.co.uk/commentisfree/cif-green/2009/dec/14/climate-change-battle-redefine-humanity.

Platt, Reg and Simon Retallack (2009) *Consumer Power: How the Public Thinks Lower-Carbon Behaviour Could Be Made Mainstream*. London: Institute for Public Policy Research.

Sharpe, Matthew and Geoff M. Boucher (2010) *Žižek and Politics: A Critical Introduction*. Edinburgh: Edinburgh University Press.

Swyngedouw, Erik (2013) The non-political politics of climate change, *ACME: An International E-Journal for Critical Geographies*, 12 (1): 1–8.

Varul, Matthias Zick and Dana Wilson-Kovacs (2008) *Fair Trade Consumerism as an Everyday Ethical Practice – A Comparative Perspective*, ESRC/University of Exeter, http://people.exeter.ac.uk/mzv201/FT%20Results.pdf.

Žižek, Slavoj (2000) *The Ticklish Subject: The Absent Centre of Political Ontology*. London: Verso.

Žižek, Slavoj (2008) *In Defense of Lost Causes*. London: Verso.

INDEX